Sleep

SLEEP

J. Allan Hobson

SCIENTIFIC AMERICAN LIBRARY

A division of HPHLP
New York

Library of Congress Cataloging-in-Publication Data

Hobson, J. Allan, 1933–
   Sleep/J. Allan Hobson.
       p.   cm.
   Bibliography: p.
   Includes index.
   ISBN 0-7167-5050-3
   ISBN 0-7167-6014-2 (pbk)
   1. Sleep.   I. Title.
QP425.H555   1989
612'.821—dc19   88-37027
                                          CIP

ISSN 1040-3213

Printed in the United States of America.

Scientific American Library
A Division of HPHLP
New York

Distributed by W. H. Freeman and Company
41 Madison Avenue, New York, New York 10010 and
20 Beaumont Street, Oxford OXI 2NQ, England

First printing 1995, HAW

This book is number 27 of a series.

To Edward Evarts and Frederick Snyder

Sleep is supposed to be,
By souls of sanity,
The shutting of an eye.

Emily Dickinson

Contents

Preface

I began planning *Sleep* while finishing my first book, *The Dreaming Brain*. Having focused narrowly on dreaming, I felt that another book was needed, one that would look at the field of sleep research in all its glorious interdisciplinary breadth. I had earlier been both flattered and tempted when Linda Chaput had suggested that I publish *The Dreaming Brain* as a volume of the Scientific American Library. Although that didn't happen, the series seemed like an ideal home for a broader view of sleep that could be enhanced by the superb illustrations that distinguish the series. I thank Linda for her patient confidence in me, and I thank her staff at W. H. Freeman and Company for their enthusiastic collaboration in its production.

The actual writing of the book was an experiment in intensity. I produced my first, longhand draft of 475 pages in only five weeks at the Villa Serbelloni in Bellagio, Italy, during a stay generously supported by the Rockefeller Foundation. In my studio, a cell called Santa Caterina, I became better acquainted with the work of many distinguished colleagues described to me by the 200 reprints that I had carried over in an old wine carton. I especially thank Truett Allison, Wilse Webb, and Anthony Kales for being such good company during my confinement. Although they and others inspired me, the ideas that flowed from my fountain pen during that rainy October in 1987 are my own responsibility.

In order to make my story coherent, I decided to omit much detail and documentation, and even to temper my usual persnickety critical reserve. I wrote in one voice, almost in one breath. I hope that my peers will not be too offended by my effort to make so many difficult scientific matters both accessible and exciting. Gentle reader, please tell me whether or not my personal approach works for you. Insofar as the message is clear, and plainly put, I thank Susan Moran and Randee Falk, who queried every sentence in the book and patiently queried again when I became temporarily blind to their little yellow flags of caution!

Perhaps the best way to thank all of the other people who have contributed to this book is to acknowledge, with strong emphasis, the personal and scientific freedom that I have enjoyed. My family, my friends, my university, my hospital, my colleagues, my sponsors, and my patients have all been so very supportive and so very, very tolerant. I hope my use of that freedom has served the larger interests of experiment and of truth.

I have dedicated *Sleep* to the two National Institutes of Health scientists who introduced me to the study of sleep and dreaming. Both have recently died, and death took both of them before their time. I am very sorry that I will not have their reaction to this book, and I hope that their memory will not be ill-served by my dedicating it to them.

Frederick Snyder was a naturalist through and through. He loved animals and felt they had much to teach us. How right he was: even our understanding of dreaming, that quintessentially human trait, is improved by a close study of our mammalian forebears. The corollary of Fred's naturalistic bent was his discomfort with human pretensions and petty conflicts. He once said, "We have enough trouble understanding nature without fighting with each other." He quit his own psychoanalysis in disappointment with both doctrine and doctor, and he left the NIH to become a public health psychiatrist to Native Americans. His scientific legacy is his conception of REM sleep as a third organismic state and his conviction that the study of evolution would help us understand the function of sleep.

Edward Evarts saw clearly the connection between neurobiology and psychiatry. He developed ingenious techniques for studying the electrical activity of individual neurons in active cats and in monkeys, and he devised accurate ways of analyzing the millions of action potentials that streamed from his subject's brains. On the basis of his findings, Evarts articulated the first specific theory of the neurophysiological difference between waking and sleep. Ed's guiding idea was that by studying sleep we could begin our difficult assault on the biology of mental illness. He once said, "If we cannot understand such global and major changes in the state of the brain-mind as occur in sleep, what chance will we have with the far more subtle changes in consciousness that trouble our patients." In his scientific work, Ed Evarts' asceticism gave rise to a classical spareness that hid from all but a few close friends his deeply passionate—and richly humorous—response to life.

Ed Evarts and Fred Snyder were both psychiatrists who were self-critical enough to be dissatisfied with the scientific status of their field and, at the same time, bold, patient, and hopeful enough to do something creative about it.

J. Allan Hobson   *1988*

Preparing this paperback edition has given me the chance to reflect upon the important changes in sleep science that have occurred between 1988 and 1995. I am grateful to Dr. Lia Silvestri of the University of Messina for providing me with ideal working conditions for this pleasant task.

Public awareness about sleep and sleep disorders has grown exponentially in recent years. Sleep clinics are now thriving in every major U.S. city and Sleep Medicine is fast becoming a mainstream subspecialty. Specific therapeutic relief for such impairing and even life-threatening problems as sleep apnea (a respiratory sleep disorder) is now available.

As one example of how much the clinical landscape has changed, it is now clear that the disorder of irresistible sleepiness known as narcolepsy, once thought to be a psychological disorder treatable by psychotherapy, has strong genetic underpinnings. Molecular biology has likewise greatly enriched our understanding of daily biological rhythms through the development of "clock mutant" animals—including fruitflies—with behavioral timing systems quite different from normal.

The biological revolution in psychiatry has been propelled in part by the development of sleep and dream science, and the study of the brain's responses to drugs has in turn yielded important new data about sleep. For example, Prozac and other widely used serotonin receptacle-blocker drugs not only relieve depressive and obsessional states, but produce an increase in eye movement activity and dreaming in non-REM sleep. We are likely to learn more and more from the actions of these new drugs that affect the brainstem chemical control systems.

Progress has also been impressive in linking sleep to bodily health via the immune system and hormone release. Sleep may thus be construed as the body's own best state of defense against infection and its most productive state of growth and development. Because almost all human growth-hormone output from the pituitary gland occurs during deep non-REM sleep, it may even be possible to boost production of that important hormone by pharmacologically increasing sleep.

The control of REM sleep—with its attendant developmental and maintenance functions—has recently been shown to be regulated over the very long term by a specific region of the brainstem. This discovery is the equivalent of finding the brain's regulatory regions for food intake or sex hormone release and provides new experimental and therapeutic opportunities to turn sleep on and off at will or even to develop biochemical substitutes for it.

The development of home-based sleep monitoring systems, which permit much more extensive, naturalistic, and cost-effective data collection than was previously possible, is also revolutionizing scientific dream research. The richness and strangeness of children's dreams, the similar profiles of dream emotions in young men and women, and the dramatic discontinuity of some dream plots have been documented and quantified. Dream theory is now clearly a scientific domain and far more secure—and far more interesting—than it was in the long era of psychological speculation.

More broadly, the growth of scientific interest in dreaming and sleep is part of a great upsurge in scientific interest in consciousness itself. The dark age of positivist and behaviorist prohibition against study of mental processes has given way to modern neuroscience and the development of imaging techniques for studying the human brain in action. And the actions which can now be "imaged" include sleep and dreaming. For the first time in human history, after millennia of philosophical rumination, we have a chance to develop a scientific theory of consciousness.

J. Allan Hobson

April, 1995

Sleep

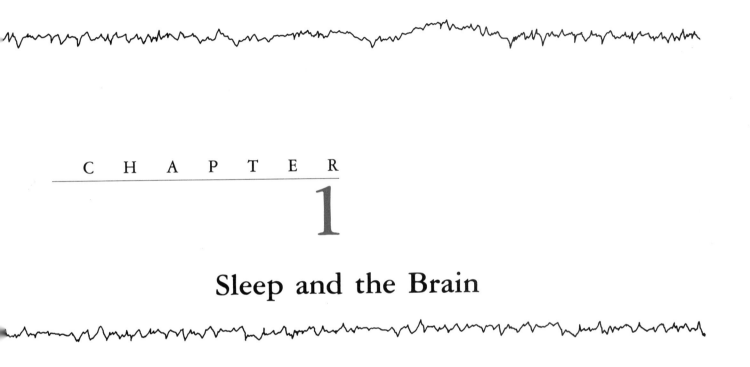

1

Sleep and the Brain

More has been learned about sleep in the past 60 years than in the preceding 6000. In this short period of time, researchers have discovered that sleep is a dynamic behavior. Not simply the absence of waking, sleep is a special activity of the brain, controlled by elaborate and precise mechanisms. Not simply a state of rest, sleep has its own specific, positive functions. The recent discoveries about sleep have been made possible by the growth of brain science. In turn, sleep research, as a leading edge of that fast-growing field, is already being used to explore the physical basis of human consciousness.

The study of sleep and especially of dreaming is changing humankind's view of itself by shedding light on the ways in which our daily behavioral rhythms, our perceptions, our feelings, and our thoughts reflect the detailed workings of the 100 billion nerve cells within our head that spark, secrete juices, and code and store data in continuous harmony around the clock. By

Modern sleep and dream research takes advantage of the fact that all mammals have the same fundamental brain structures and functions, as can be seen in these two similar brain wave patterns recorded from a man and a cat during REM sleep. Thus, the most basic questions about sleep and dreaming can be answered by neurobiological experiments whose results are used to shape behavioral and psychological theories.

Sleep, as we mammals know it, is much more complex than the states of bodily rest we share with these heliconia butterflies and all other living creatures. The differences between sleep and rest are functions of the brain.

understanding how the brain's incessant spontaneous activity is rendered both intrinsically orderly and finely tuned to external realities, we may someday even begin to glimpse the biological basis of those most glorious of all human qualities, imagination and creativity.

Sleep, as I will use the term, is found only in animals with highly developed brains. In sleep the brain remains active, but it does not effectively process information received from the senses. Since less highly evolved animals never stop reacting to the outside world, these animals rest but do not truly sleep. The central role of the brain in sleep can be conveyed by rephrasing Abraham Lincoln's famous declaration about government: sleep is of the brain, by the brain, and for the brain. This is not to say that no other part of the body participates in or benefits from sleep. But it is to emphasize that for sleep to occur a highly developed brain is necessary.

OF THE BRAIN

We say that sleep is of the brain not just because it is fully developed only in animals having elaborate brains, but because the brain itself is the seat of the most impressive changes in sleep. The behavioral symptoms of sleep—a lying down posture, closed eyes, and lack of responsiveness to stimulation—can easily be feigned. Impossible to feign are the brain signs of sleep. Measurable changes in the electrical activity of the brain distinctively define sleep.

By measuring electrical activity, we are able to distinguish sleep from its unconscious imitation by other behaviors. The hibernating woodchuck outwardly appears to be sleeping. But the electrical activity of its brain has virtually ceased—something that never occurs in sleep. Thus, sleep is an active state of the brain: brain electrical activity continues throughout sleep and differs from that during waking. Sleep is of the brain.

BY THE BRAIN

We know now, too, that the brain controls itself so as to *produce* sleep. The clocks that turn sleep on and off are composed of networks of brain cells. These clocks not only time whether we sleep or wake but also program an elaborate and orderly sequence of brain events within sleep. In one such event, the continuously active brain becomes extraordinarily more active every 90 to 100 minutes during sleep and remains so as long as an hour. It is during such periods that we dream. Thus, dreaming, like sleep, is of the brain and by the brain.

We are used to thinking of such conscious mental activity as our remembered dreams as being the product of the "mind." To say that dreaming is of the brain is not to say that dreaming is not of the mind—or even, in some restricted sense, by the mind. It is rather to emphasize that every state of mind

A defining feature of mammalian sleep is a distinctive change in the electrical activity of the brain (left panel). The brain mechanisms regulating sleep (center panel) involve populations of neurons (represented by the green diamond-shaped symbols) that change their chemical properties during sleep in such a way as to benefit subsequent waking (right panel).

Of the Brain

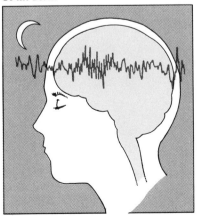

The brain's own electrical activity changes...

By the Brain

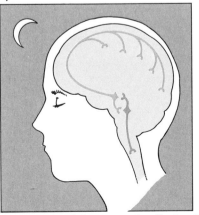

in response to signals from networks of brain cells...

For the Brain

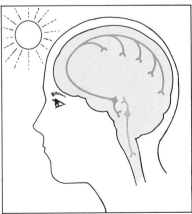

with resulting improved function.

reflects a state of brain. We can therefore now make bold to say that dreaming is caused by brain activation in sleep. Not by indigestion, not by train whistles, not by angels, and not by wishes. Sleep and dreaming are by the brain.

FOR THE BRAIN

The brain is the prime beneficiary of sleep, as is made obvious by the progressive decline in our cerebral capacities when we are deprived of sleep. We first have difficulty concentrating, attending, and performing coordinated motor acts such as driving cars, then we become irritable and suffer an almost painful sleepiness. Taken to an extreme in "brainwashing," sleep deprivation can cause even heroically patriotic citizens to denounce their own nations and ideals, to sign patently false declarations, and to join political movements that have been lifelong anathemas to them.

After we go five to ten days without sleep, our brain loses its bearings altogether and madness takes over: the trusting become paranoid; the rational, irrational; and the sane begin to see and hear things that aren't there. All the dysfunctions caused by sleep deprivation are rapidly reversed when lost sleep is recovered. We don't yet know exactly how sleep ensures efficient brain function, but that it does so is beyond doubt. Thus, sleep is for the brain. Sleep is of the brain, by the brain, and for the brain.

THE HISTORICAL NEGLECT OF SLEEP

Now that we recognize these three principles, we can better appreciate why sleep science is so new. Until recently, most naturalists did not recognize that sleep involved brain activity, and even though as far back as classical Greek and Roman times, certain exceptional philosophers and naturalists reasoned that both sleep and dreaming were affairs of the brain, they had no means of proving it. It was not until the second quarter of our own century that a way was found to measure sleep objectively by recording the brain electrical activity.

Without an objective means of studying sleep, the best that even the most diligent observers could do was to watch. And early observations of sleep did not proceed far, for several reasons. One is purely pragmatic. Humans are used to sleeping at certain times, so the observer's brain tends to go out of commission at the same time as that of its subject. An observer whose own brain wants to sleep won't watch the sleep of another subject for long.

Many animals sleep when our own brains are alert, however. Why, then, do we have so few good descriptions of animal sleep? The reason is simple: the outward appearance of a sleeping animal is by and large so tranquil that it was commonly assumed that there was little to be observed. And because their own consciousness was obliterated in sleep, observers wrongly concluded that there was little going on under the tranquil surface of the sleeping animal.

In retrospect, two factors should have combined to produce an observational sleep science much earlier than the twentieth century. First, the subjective experience of nightmares and dreams should have dissuaded those who believed sleep could occur without intense brain activity. Second, while animal sleep appears tranquil, in all mammals—including humans—observable movements of the eyes, face, and fingers occur periodically during sleep. It seems inconceivable to us that those who watched sleeping cats' paws twitch, sleeping puppies' feet scamper, or sleeping babies' faces grimace and smile did not suspect intense underlying brain activity. Indeed, both the Roman naturalist Lucretius and the seventeenth-century Italian scientist Lucio Fontana made such observations of mammals in sleep. They even went so far as to infer correctly that such outward signs of motion were related to the inward experience of dreams. But they did not recognize that this inference was in turn strong presumptive evidence of brain activation.

EARLY BIOLOGICAL STUDIES

The undervaluation of sleep was continued by early sleep researchers, who mistakenly thought their studies showed that only waking was an active state of the brain. In 1809, the Italian physiologist Luigi Rolando noted the effects of removing the frontal parts of the brain in birds: a hen, a cock, and several ducks and hawks all appeared sleepier after this intervention. Rolando concluded that the brain is the seat of wakefulness. This conclusion was echoed by the French physiologist Marie-Jean-Pierre Flourens, who in the years 1814 through 1822 observed the effects of removing from pigeons the main frontal structures, or cerebral hemispheres. These studies strengthened the idea that sleep was a passive process, the result of simply the subtraction of waking.

The nineteenth-century Czechoslovakian physiologist Jan-Evangelista Purkinje advanced the argument that some activities based in the upper parts of the brain could not be maintained without the influence of lower brain centers. (The upper parts of the brain in humans, whose posture is vertical, correspond to the frontal parts of the brain in other animals.) In doing so, he was hinting that the lower brain may activate the upper brain and so produce the wake state. But Purkinje was also inclined to believe that only waking—and not sleep—was an active state of the brain, with its own mechanism and its own function.

MISSED OPPORTUNITIES

The nineteenth century also saw the introduction of several mechanical techniques that could have facilitated the objective observation of sleep. Foremost among these was photography. The photographic technology of Louis Jaques Mandé Daguerre and later that of the Lumière brothers made possible the automatic, quasi-permanent recording of visual data. The Englishman Eadweard Muybridge created the bridge to cinema with his striking time-lapse studies of animal motion. In one of his first attempts, he was employed by Leland Stanford to prove photographically that a racehorse often has all four feet off the ground. He photographed a running horse at short, regular intervals, then reassembled these sequences of photographs as a series on a disc. When viewed through a slot as they whirled around, the photographs appeared to show movement. And, indeed, all four legs did leave the ground. One of Muybridge's series appears at the bottom of this page.

But, curiously, these newly available mechanical techniques were not applied to the study of sleep. Despite the unavailability of electronic timers, it would have been quite easy to take photographs of sleeping humans at intervals as short as one minute. Such photographs would have revealed the periodic activation during sleep of the motor centers of the brain caused by movements and established at once that sleep behavior is both dynamic and rhythmic.

Why were these methods not applied? A key reason is that, as in earlier periods, no one anticipated that sleep observation could produce a fruitful result. Sleep seemed like such an inert state that static postures were taken for granted. And the conviction of many, including the great English neurophysiologist Sir Charles Sherrington and the Russian behavioral physiologist Ivan

Eadweard Muybridge, who convincingly demonstrates here that running horses fly, must have documented almost every behavior except sleep with his exquisite time-lapse photograph sequences. With his techniques, the posture shifts that demarcate sleep and even the smaller movements of face and eyes that accompany dreams could have been objectified almost a century ago.

Animal motion, a subject of interest to many naturalists and physiologists, was observed and recorded in new ways. For example, the Frenchman Etienne-Jules Marey was able to record repetitious aspects of motile behavior, such as the wing beats of a pigeon in flight, by making use of moving smoked drums. Today, sleep laboratory polygraphs record the electrical activity of the brain as well as movements of the eyes and body.

Pavlov, was that the brain cells shut off in sleep. We now know that most brain cells decrease their activity by only 10 to 20 percent, even in the deepest stages of sleep.

A second factor also discouraged the photographic study of sleep. Suggest such a study to most people and they politely but firmly decline. Sleep is very private, and the bedroom is an intimate, almost sacred place. Moreover, we tend to feel that in sleep we are in every way unprotected and thus unsafe. For these reasons, there is a strong taboo against observing, and especially photographing, sleep.

Another trend that worked against sleep studies was the rise of the psychological movement called behaviorism. Beginning with Pavlov, psychologists became increasingly preoccupied with discrete motor acts and consequently ignored the background states out of which such acts arose. B. F. Skinner went so far as to suggest that the brain was a "black box," of no interest or relevance to psychology.

THE BRAINSTEM: A CENTER FOR SLEEP AND WAKING

Beginning in the late nineteenth century, scientists began to confirm and extend the view hinted at by Purkinje's work that the activity of the lower brain keeps us awake. A series of clinical observations, culminating during the frightful influenza epidemic of 1918, forced scientists to entertain the idea

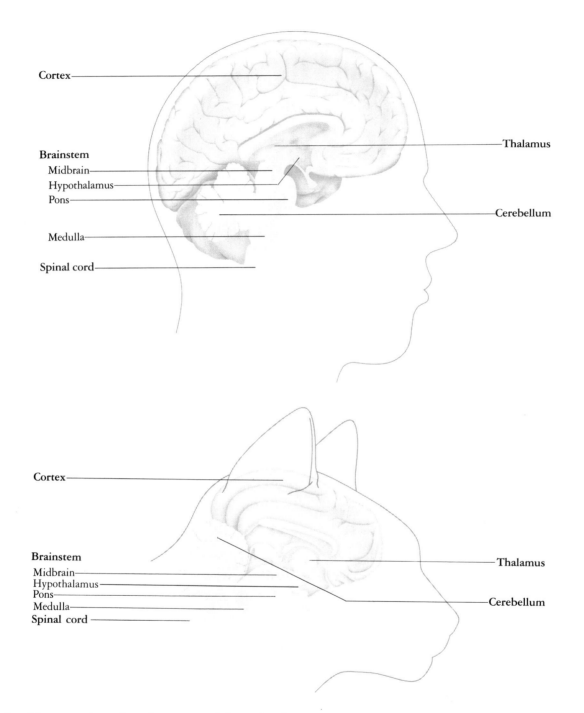

Cortex

Brainstem
Midbrain
Hypothalamus
Pons

Medulla

Spinal cord

Thalamus

Cerebellum

Cortex

Brainstem
Midbrain
Hypothalamus
Pons
Medulla
Spinal cord

Thalamus

Cerebellum

The brains of humans and cats share the same general design, but the size and shape of the parts differ. In humans the cerebral cortex and thalamus are not only larger but much more complicated, reflecting our greater cognitive capacity. The cat's brainstem and cerebellum are relatively larger than the human's but have the same degree of complexity, reflecting the similarity of such universal mammalian behaviors as sleep. Because the brainstems in cats and humans are so similar, the cat is a good subject for neurophysiological studies of sleep.

that both sleep and waking might be actively controlled by a brain structure called the brainstem.

The brainstem is the enlarged upward extension of the spinal cord. It gives rise, like the calyx of a flower, to the cerebral hemispheres—consisting of the thalamus and cortex. The brainstem connects these "higher" structures to the spinal cord, with its many nerve roots carrying impulses to and from the skin and muscles below. This botanical analogy is particularly apt because the brainstem is the seat of mechanisms that control visceral functions like body temperature, sexuality, and appetite that have been called "vegetative." Incredibly complex, the brainstem also contains not only sensory and motor nerves of the head, face, and eyes, but also the nervous elements that coordinate body position, muscle tone, level of arousal, and cardiorespiratory functions, among others. As shown in the diagram on the opposite page, the brainstem consists of the medulla and pons, the midbrain, and the hypothalamus.

In 1875 Alphonse Gayet, the chief of surgery at a hospital in Lyon, France, found that the brainstem (specifically, a large part of the midbrain) had been damaged in a patient whose eyes were paralyzed and who was lethargic although not comatose. The paralysis of the eyes was expected; however, the lethargy came as a surprise. Gayet concluded that sensory messages from the eyes and other organs were necessary to maintain wakefulness. He thought that the blockage of these messages explained his patient's lethargy and so failed to recognize that the brainstem might support wakefulness without having to rely on sensory stimulation.

That insight, as well as the companion hypothesis that the brainstem supports sleep, was first articulated by the Viennese neurologist Constantin von Economo. Following infection by the influenza virus, von Economo's patients displayed two distinct kinds of pathology: insomnia with overactivity or excessive sleepiness. Because patients with insomnia and overactivity had damage to the front of the midbrain, while those with excessive sleepiness or encephalitis lethargica had damage to the back of the midbrain, von Economo postulated that the brainstem contained one system to regulate waking and another to regulate sleep—that is, a waking center and a sleep center. Von Economo further speculated that the two centers worked via chemical substances.

The idea that chemical substances played a role had been generally accepted since the early work of Henri Pieron, the Parisian psychologist and physiologist, at the beginning of this century. In fact, the notion of a sleep substance that might circulate in the blood or spinal fluid grew naturally out of the nineteenth-century metabolic physiology. Muscle fatigue and the breakdown of stored energy sources were known to produce specific chemicals, which circulated in body fluids until excreted by the kidneys or further pro-

Constantin von Economo was a volunteer lieutenant in the Austrian Army Air Force from 1915 to 1916. He observed that influenza left many patients whose brains had been infected by the virus unable to stay awake. Von Economo observed that this condition (called encephalitis lethargica) was associated with damage to neurons in the brainstem. He correctly inferred that normal waking was actively mediated by the brain.

cessed by the liver. It was natural to suppose—and many scientists still do—that such molecules would accumulate over the day and exert an increasing soporific effect on the brain, ultimately causing sleep.

To test this idea, Pieron transferred spinal fluid from sleep-deprived dogs into wakeful dogs and noted their ensuing behavior. Since that behavior was often sleeplike, Pieron postulated the existence of a sleep substance or sleep factor and began a tradition that remains vigorous and full of promise to this day.

The state of affairs in sleep research when the field's modern era began in the 1920s thus clearly reflected trends that can be traced back to as early as 1800 but that crystallized between 1890 and 1920. Neurologists knew that the brain was important and that it might even contain specific mechanisms for regulating behavior and consciousness. During the last half of the nineteenth century, speculations regarding the brain basis of consciousness were particularly rife in Germany and Austria. At that time, the neurophysiologist Hermann Helmholtz and his pupil Wilhelm Wundt developed the new field of experimental psychology, which attempted to test psychological principles in the laboratory.

These currents emerged in the comprehensive vision of William James, whose *The Principles of Psychology* was published in 1890. According to James, brain science, introspective self-observation, and experimental psychology were all to be integrated in one discipline. But as these three special fields progressed in the first quarter of the twentieth century, each of them tended to diverge from, and to denigrate, the others. Thus, the study of dreaming had by the 1920s become the province of psychoanalysis, a branch of psychology that eschewed neurology and the rest of the scientific establishment. The behaviorist mainstream of experimental psychology gave up dreaming—and all other forms of mental experience—as hopelessly subjective. The behaviorists ultimately would come to eschew neurology as vigorously as the psychoanalysts had done, but for even more radical reasons: while Freud feared a weakening of his new theory and his burgeoning psychoanalytic movement by the neurology he believed might ultimately replace them, the leading behaviorist, B. F. Skinner, flatly denied that neurology could ever be useful to psychology. This was the state of affairs in 1950.

The subsequent developments in sleep research, which are the main subject of this book, have led to changes that are still in progress and that may well produce a reasonable balance between studies of brain (neurology), of mind (psychology), and of behavior (ethology). If so, we should be able to make use of, and actively integrate, the findings of all three fields. We now look with great interest to neurophysiology, which can help us understand what the brain does in sleep, how it does it, and ultimately, why it does it; we look to psychology for the definition and measurement of mental processes during sleep, how they differ from those of waking, and what benefit the waking mind may derive from sleep; and we look to ethology for objective evidence of the outward manifestations of sleep, for their relationship to such wake-state survival behaviors as feeding and reproduction, and to such information-processing activities as attention and learning, and for ways to understand their adaptive or ecological significance.

RECORDING SIGNALS FROM THE BRAIN

The first phase of modern sleep research, which lasted from 1928 to 1953, saw the beginning of a quantum leap in our knowledge about sleep. This leap in knowledge occurred for one major reason: scientists finally had the ability to study the brain as a net of neurons whose individual and cooperative activity has measurable electrical and chemical properties. The modern concept of the brain as a neuronal colony derives from the work of Santiago Ramón y Cajal, a neuroanatomist who discovered in 1890 that the neuron was the structural and functional unit of the brain.

The brain contains billions of cells called neurons, which communicate with one another via electrical and chemical signals. Each neuron in the brain concentrates an electrical charge across its surrounding membrane; hence, each neuron is a power source. When the neuron is sufficiently excited by a chemical message from its neighbors, charged particles, called ions, suddenly rush across the semipermeable membrane of the neuron, causing the voltage to

The India ink drawing of the mouse brainstem was made by Ramón y Cajal. It illustrates the complex meshwork of cells and fibers called the reticular formation, which runs down the center of the brainstem. In the orange box (labeled N by Ramón y Cajal) and in the photograph on the right are two giant pontine reticular neurons of the type that become active in the dreaming phase of sleep.

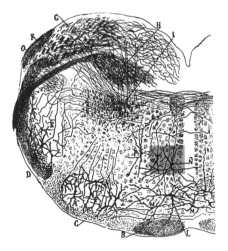

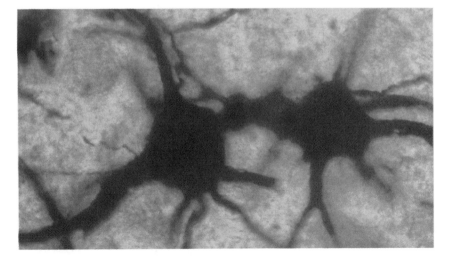

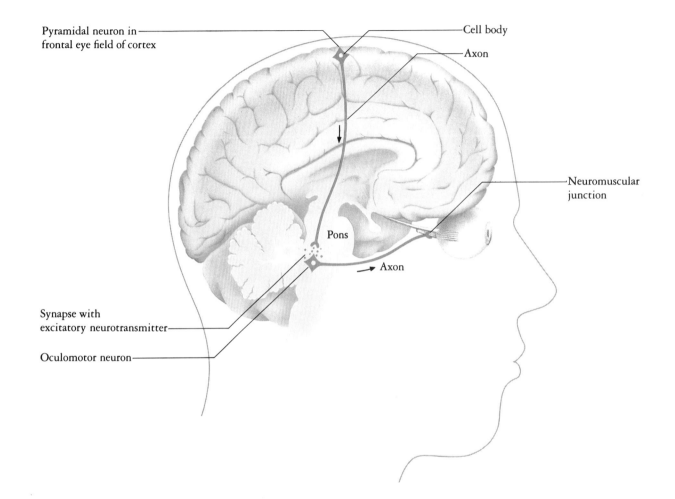

Pyramidal neuron in
frontal eye field of cortex

Cell body

Axon

Neuromuscular
junction

Pons

Axon

Synapse with
excitatory neurotransmitter

Oculomotor neuron

When we voluntarily shift our gaze, neurons in the frontal eye field of our cortex send action potentials down their axons to the pons in the brainstem. There they release neurotransmitters that excite oculomotor neurons, which relay the movement command to the appropriate eye muscles.

change dramatically. Scientists say the cell "fires" or "discharges." The change in voltage is called the action potential; it spreads from the cell body down the axon, a wirelike fiber that ends at a junction with another cell. At the junction, or synapse, the electrical charge causes one or more chemicals, called neurotransmitters, to be released. These neurotransmitters in turn increase (if they are excitatory) or decrease (if they are inhibitory) the probability that neighboring cells will generate more action potentials. Action potential generation goes on continuously night and day, with some neurons firing upward of 100 times per second. The interconnections between many neurons form a chain of communication called a circuit. The diagram on this page shows part of one of these circuits, that commanding eye movement.

Technical advances in electronics soon allowed scientists to measure the electrical activity of the brain and its constituent neurons. In 1928, the German psychiatrist Hans Berger successfully recorded continuous electrical activity from the scalp of human subjects. Berger correctly inferred that the signals he recorded, which he called electroencephalograms (or EEGs), were of brain origin. Such electrical signals had long been known to be detectable, but only from the surface of animal brains. The way that Berger convinced himself—and his many skeptical colleagues—that the signals he registered were indeed of brain origin (and not from skin, muscle, or movement) was by noting that when subjects closed their eyes, became drowsy, and went to sleep, there were distinctive changes from the pattern initially observed.

When two electrodes are placed on the scalp and connected to an amplifier-recorder (the EEG machine), the voltage between the two can be recorded continuously. A record of the EEG is made by moving paper at a constant speed under a pen whose vertical position is determined by the voltage level of the brain. The example on this page shows an EEG tracing and illustrates two of its significant features, frequency and amplitude. Although the exact source of the EEG in the brain is still not known, each blip or wiggle in the record has some as yet to be precisely defined relationship to the activity of the many, many neurons in the vicinity of the two electrodes.

In a sleep laboratory, electrodes are attached to the subject's head so that brain, muscle, and eye activity can be monitored on a polygraph in the next room. One of the cardinal brain signs of sleep, the Stage II EEG spindle, is shown in the tracing on the right.

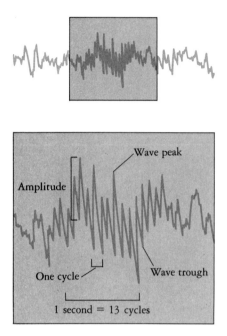

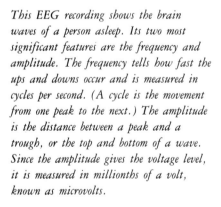

This EEG recording shows the brain waves of a person asleep. Its two most significant features are the frequency and amplitude. The frequency tells how fast the ups and downs occur and is measured in cycles per second. (A cycle is the movement from one peak to the next.) The amplitude is the distance between a peak and a trough, or the top and bottom of a wave. Since the amplitude gives the voltage level, it is measured in millionths of a volt, known as microvolts.

What Berger first obtained, when subjects were awake and had their eyes open, was a very fast, low-voltage EEG. Such a pattern is indeed characteristic of arousal. The high frequency probably reflects the high level of activity, that is, the high level of discharge of many cells. It has recently been shown that during waking the activity of neurons and of the corresponding brain waves is rhythmic at 40 cycles per second. The low voltage is recorded because the discharge is desynchronized, or arhythmic. A desynchronized wave form indicates that the action potentials responsible for the signal occur at different instants in time.

When Berger's subjects closed their eyes or dozed, their EEGs showed a progressive shift toward synchronization, that is, toward a slower, higher-voltage EEG. The term *synchronization* was used because the firing of neurons seemed more coordinated, and more fired simultaneously, leading to larger net voltages. Yet the frequency was lower because net activity had actually declined.

Merely closing their eyes, and thus relieving their brains of visual stimulation, shifted the EEG pattern of Berger's patients from the desynchronized, low-voltage (less than 50 microvolts), fast (15 to 50 cycles per second) pattern to a regular wave form of higher amplitude, known as alpha (8 to 12 cycles per second). This shift, which has become almost sacred to practitioners of transcendental meditation or systematic relaxation, demonstrates the utility of the EEG as a brain state detector and reveals the sensitivity of the brain to major changes in input. The shift to higher-voltage, slower EEG rhythms (increasing synchronization) continued as Berger's subjects began to actually doze. These findings seemed to confirm what many had thought—that waking was dependent on sensory stimulation and that sleep was the simple consequence of decreasing the stimulus level. Both assumptions were later proved to be incorrect.

THE EEG OF SLEEP

The response time of Berger's mechanical EEG machine was limited by the physical inertia of the system to tenths of a second. To record the action potentials of neurons a faster system was required, and this need was met in 1933 by Edgar Adrian and Brian Matthews' oscilloscope. The oscilloscope could respond in thousandths of a second, because its pen was a beam of virtually weightless electrons and its paper a fluorescent screen that glowed when struck by the passing electrons. These fast-fading traces, being luminous, could be photographed. The electron beam was swept back and forth by an electronic circuit, and the neuronal signals caused it to jump up and down by changing the power of electromagnets above and below the electron path.

Adrian used this new device not only to confirm Berger's findings but to embark on several new lines of research, including the study of animal sleep. In studying the EEG of several species, Adrian found that odors as well as light had powerful effects on the wave forms. In the hedgehog, for example, a characteristic EEG pattern, which Adrian called the olfactory-induced wave, spread over the cortex each time the animal, in drawing a breath, whiffed odorants across its olfactory nerve endings. This was an example of what is now called an evoked potential, the EEG response to sensory stimulation. Both the spontaneous and evoked EEG potentials changed dramatically in sleep.

By the 1930s, the EEG was a well-established instrument of investigation, and Berger's findings had been amply confirmed and extended. The tendency for the EEG to slow in frequency and to grow in amplitude proved to be continuous as sleep continued and deepened. Thus, when human subjects passed from arousal to relaxed waking, to drowsiness, to light sleep, and finally to deep sleep, their brain waves changed in a characteristic way. The characteristic patterns seen from light sleep to deep sleep have become known as the four stages of sleep; the distinguishing features of the waves are given in the following table:

Characteristics of EEG sleep stages

Stage	Frequency (cycles per second)	Amplitude (microvolts)	Wave form
Stage I	4–8	50–100	Theta waves
Stage II	8–15	50–150	Spindle waves
Stage III	2–4	100–150	Slow waves plus spindles
Stage IV	0.5–2	100–200	Delta waves

The EEG patterns of the four sleep stages and waking are shown in the figure on page 16. In this progression of EEG stages, the frequency decreases (we say the EEG slows), and, at the same time, the amplitude increases (we say the EEG is of higher voltage). Although the progression of EEG waves has been divided into discrete stages, it is actually gradual and continuous. Once Stage IV is reached, the process reverses itself, and the sleeper goes back through Stages II and III to Stage I. This now clearly cyclical process tends to repeat itself, from Stage I to Stage I, at 90- to 100-minute intervals throughout the night.

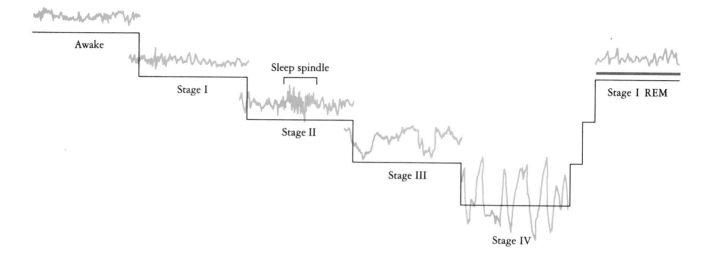

After sleep onset, the EEG changes progressively from a pattern of low voltage and fast frequency to one of high voltage and low frequency. The four stages of non-REM sleep shown above are successive steps in this process, which occupies the first 50 to 70 minutes of sleep. The EEG is then reactivated to the low-voltage, fast condition of Stage I REM, the phase of sleep when dreams occur (indicated by the purple bar).

The first definitive EEG sign of sleep occurs in Stage II, when a complex wave sequence called a sleep spindle is seen in the EEG recording. In a sleep spindle, first each successive wave increases in amplitude and then, after the amplitude reaches a peak, each successive wave decreases in amplitude.

Until 1953, the EEG patterns during sleep were thought to be uniformly deactivated compared with those of waking; that is, the voltage of the waves was higher and the frequency distinctly slower. It thus came as an almost unbelievable surprise to find that recurring periodically in sleep were phases of EEG activation with low-voltage, fast waves that were as intense as those seen in waking. These activated EEG patterns were most pronounced during Stage I of the sleep cycle.

The periods of activation were discovered by Eugene Aserinsky, working in the Chicago laboratory of Nathaniel Kleitman. Aserinsky's original interest had been in eye movements as an index of arousal in children, and he found, to his surprise, that the periods of EEG activation that occurred in sleep were frequently associated with flurries of rapid eye movements (now called REMs), as well as with increases in pulse and respiratory rate. These periods are now called Stage I–REM sleep, and the graded stages I through IV of slow-wave, synchronized EEG patterns are now called non-REM sleep.

Kleitman immediately intuited that this physiological activation of the brain was likely associated with the psychological activation of dreaming. He was proved correct by later laboratory studies in which subjects gave reports following awakenings from non-REM and REM sleep. Because the laboratory tends to inhibit dream content, many subsequent studies have relied on home-based reports—and drawings—recorded in journals.

THE CONTROL OF SLEEP BY THE BRAINSTEM

By the 1950s, the nature of sleep and waking were more clearly understood. The view of waking as dependent on sensory stimuli and the view of sleep as the absence of waking and of activity had been disproved. In one of several studies providing the critical evidence, the Swiss physiologist Walter R. Hess showed that sleep could be induced by electrically stimulating the thalamus. His results implied that sleep was caused by a change in the pattern of neuronal activity and not by a cessation of that activity.

That the alternation between sleep and waking was caused by the intrinsic activity of the brainstem—and not by sensory input—was later demonstrated by a series of experiments. In the early 1930s the Belgian physiologist Frederic Bremer isolated the hypothalamus, thalamus, and cortex of a cat from the rest of the brain by transecting the midbrain. This isolated forebrain, or

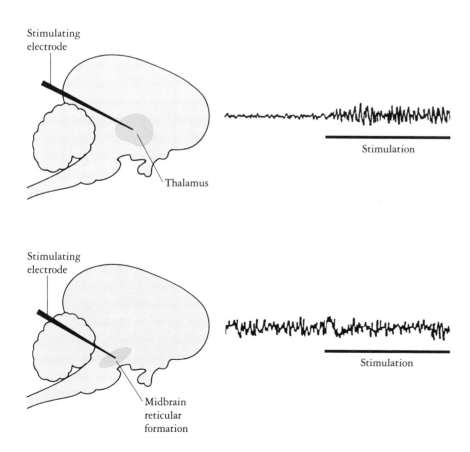

Stimulation experiments produced convincing evidence that EEG changes in sleep and waking were reflections of the brain mechanisms underlying those states. When Walter Hess stimulated the thalamus of waking cats with low-frequency pulses, the animals went to sleep and their EEGs shifted to a high-voltage, slow pattern. Conversely, when Horace Magoun and Giuseppe Moruzzi stimulated the midbrain of sleeping cats with high-frequency pulses, the animals were promptly aroused and their EEGs were activated.

cerveau isolé, produced the EEG patterns of sleep which could only be reversed by olfactory stimulation.

The crucial 1949 study by Giuseppe Moruzzi and Horace Magoun, at Northwestern University, found that high-frequency stimulation of the midbrain produced behavioral arousal and EEG desynchronization. A distinctive feature of the brainstem is the net or meshlike organization of many of its neurons that are not primarily components of sensory or motor pathways. It was this reticular formation of the brainstem that Moruzzi and Magoun stimulated to produce arousal; when it was destroyed by the American physiologist Donald Lindsley, animals were as somnolent as Bremer's cerveau isolé preparations even though their sensory and motor pathways were intact. The results of the experiments of Hess and of Moruzzi and Magoun are summarized in the diagram on page 17.

The obvious conclusion was that the difference between waking and sleep was due to the intrinsic activity of the brainstem: electrical impulses arising within the brain itself were capable of producing either state, depending on their site of origin and their frequency. Von Economo's active control model was thus vindicated, and the study of sleep moved definitively into the brain.

By 1953 sleep studies had reached the point where not only had several major misconceptions about sleep been overthrown, but a completely unexpected aspect had been revealed. It was made clear that far from being the passive responder to external stimuli, the brain had its own source of self-activating power—the spontaneous action potentials of the neurons. It was further made clear that neuronal activity never ceased, even during sleep, but was rather reorganized, and that this reorganization was also internally and automatically controlled. Finally, it became clear that the periodic and automatic self-activation of the brain in sleep was the physiological basis of dreaming.

The subsequent 35 years of sleep research have focused on three major topics: the localization of sleep control mechanisms in the basal forebrain and the brainstem; the investigation of the electrical activity of neurons in the brainstem during sleep; and the investigation of the chemical influences on the rest of the brain of brainstem neurons that turned on—or off—during sleep.

THE CONTROL OF REM AND NON-REM SLEEP

The detailed study of the brainstem as a sleep state generator was pursued in the Pisa laboratory of Giuseppe Moruzzi. Moruzzi and his co-workers transected the brain in the pons (between the midbrain and medulla), leaving the trigeminal nerve behind the cut, and thus detached from the upper brain. Since the trigeminal nerve carries the large sensory input from the head and face, the upper brain was no longer able to hear or feel, though it could still see and smell. In spite of its isolation from two major sources of sensory information, this preparation was as hypervigilant as Bremer's encephale isolé; that is, it remained perpetually awake. Thus, Moruzzi's idea that the brain activation was independent of sensory stimulation was strengthened.

The lack of sleep signs further suggested that structures behind the cut in the medulla might actively contribute to the EEG slowing of non-REM sleep. Studies by other researchers have implied that an area just next to the hypothalamus, called the basal forebrain, may play a role in controlling non-REM sleep. The generation of non-REM sleep was shown to be impeded by damaging and enhanced by stimulating this area of the brain.

A different brainstem structure, the pons, was shown to be critical for REM sleep generation by Michel Jouvet of Lyon, France, in 1962. As shown in the figure on this page, Jouvet transected the midbrain of a cat, then completely removed all the structures above the cut, except the hypothalamus. The resulting "pontine" cats had a periodically recurrent phase of rapid eye movements associated with a complete loss of muscle tone. This latter phenomenon (called postural atonia) had been shown also to characterize REM sleep in normal cats by Jouvet and François Michel in 1959; atonia was later shown to occur also in human REM sleep. When electrodes are attached to the chin muscle, potentials recordable during movement as the electromyogram (EMG) are found to be abolished in REM sleep. Because we do not move our muscles in REM sleep, we cannot express the motor acts of our dreams. Of course, during their atonic-REM periods Jouvet's cats could show no cortical EEG changes (because they had no cortex), but they did have spiking EEG waves in the pons during these periods that resembled those seen in normal cats. Jouvet proposed that there was both a clock and trigger mechanism for REM sleep in the pons. The clock was reliable because the episodes occurred in the cat at regular intervals of 30 minutes as in normal sleep; the trigger was effective because the episodes lasted for the normal duration of 6 to 8 minutes.

From this brief discussion, it should be clear that many areas of the complex brainstem are involved in the control of sleep and waking. Non-REM sleep mechanisms in the basal forebrain interact with medullary and midbrain reticular systems to produce EEG slow waves in the cortex; periodically interrupting this process is the REM sleep generator in the pons, which reactivates the brain.

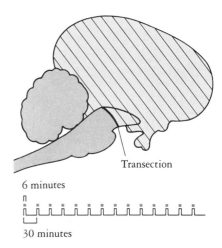

Transection

6 minutes

30 minutes

Instead of occurring only during sleep, REM activity became constantly periodic after Michel Jouvet's transections of the brainstem just anterior to the pons. This experiment pointed to a timer and trigger for REM sleep in the pons.

THE ROLE OF INHIBITION

A new technique set the stage for a description of how the individual neuron contributes to the sleep process. By inserting a fine-tipped microelectrode near a neuron, researchers could record each time the neuron fired. Using this technique, Edward Evarts performed the first recordings of well-identified and well-isolated single neurons during sleep.

When Evarts recorded the large pyramidal nerve cells of the motor cortex, which command movement via their long axons to the spinal cord, he found that they were as active in REM sleep as in waking. The pattern of firing was different, however, as shown in the figure on page 21: the neurons tended to fire in clusters of closely spaced discharges during REM sleep but regularly and one at a time during waking.

Evarts proposed in 1963 that the difference between REM sleep and waking resulted from a loss of local inhibitory control in REM. Ordinarily, when a neuron in the motor cortex fires, it is prevented from firing again too soon afterward. When the neuron fires, its axon transmits a signal to other neurons that fire in turn and transmit an inhibitory neurotransmitter back to

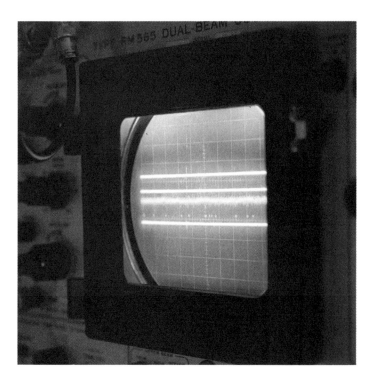

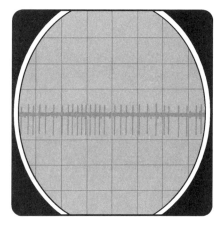

Left: *The electrical activity of individual neurons can be recorded from microelectrodes and displayed on the calibrated screen of a cathode ray oscilloscope.* Right: *Each signal (or action potential) of the neuron looks like a picket in a fence and is therefore sometimes called a "spike" by neurophysiologists.*

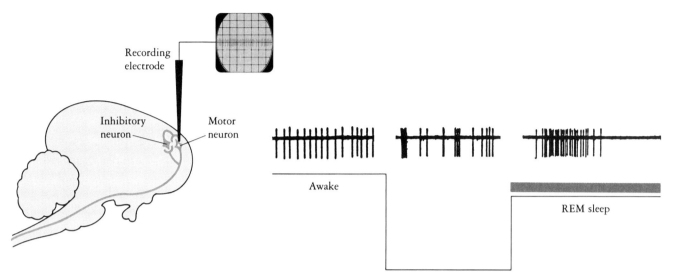

the first neuron. The inhibitory neurotransmitter prevents the first neuron from firing for a short period of time. When a neuron fired during REM sleep, however, it kept on firing instead of being damped as it was in waking. Somehow inhibition of the neuron was being prevented. This was the first specific neurophysiological hypothesis about sleep at the level of the neurons.

Left: *Motor neurons in the cortex fire regularly in waking.* Right: *The action potentials become progressively more irregular in non-REM sleep and even more so in REM sleep, when dense clusters of spikes accompany the rapid eye movements. Edward Evarts thought that this change in pattern might be due to a progressive loss of inhibition.*

THE CHEMISTRY OF SLEEP

Knowledge about the chemical nature of the brainstem neurons increased dramatically in the early 1960s when the Swedish anatomists Kjell Fuxe and Anica Dahlstrom demonstrated that the pons harbored neurons whose widely branching axons distributed two chemical substances throughout the brain. Neurons for each chemical were located in their own defined area of the pons (such an area is called a nucleus). The neurons of the midline raphe nuclei were the main source of the neurotransmitter serotonin (5-hydroxy-tryptamine), and the neurons of the nucleus locus coeruleus, a major source of the neurotransmitter norepinephrine. Both were known to be neurotransmitters that influenced the cells they contacted. Because of their widespread and diffuse distribution and the lengthy time course of their action they came to be called neuromodulators, indicating that they set the response mode of the brain rather than conveying specific sensory or motor data. Serotonin and norepinephrine belong to a class of substances called biogenic amines; hence, neurons that release these two neurotransmitters are referred to as "aminergic." Because these two substances were also known to play an important role in the control of the cardiovascular and gastrointestinal systems, they were strong candidates for a central role in controlling the state of the brain. It was as if the brain had its own nervous system, a brain within a brain as it were.

Another neurotransmitter, called acetylcholine, was also suspected to play a role in sleep. In contrast to serotonin or norepinephrine, which are often inhibitory, acetylcholine is excitatory; that is, it causes cells to fire more, not less. Neurons that release acetylcholine are called "cholinergic" and cells that respond to it are called "cholinoceptive." Having previously emphasized the role of acetylcholine in REM sleep, Jouvet proposed in 1966 that the brainstem exerted its control of non-REM sleep via serotonin and of REM sleep via norepinephrine. The idea was that each neurotransmitter caused a different state via its influence on neurons throughout the brain. These latter two conclusions, though based on solid pharmacological data, have not been confirmed by physiological or by biochemical studies, but Jouvet's earlier postulate that acetylcholine enhances REM sleep has been fully confirmed.

If, as Jouvet's work had clearly indicated, there were both a clock and a trigger for the periodic REM sleep episode in the pons, it should be possible to identify specific neurons in both. Using the microelectrode technique, Robert McCarley and I found that some cells turned on and others turned off in REM sleep. The REM-off cells were the real surprise, because they were the very neurons in the raphe and locus coeruleus that Jouvet's pharmacological studies had predicted should be on! The REM-on cells did *not* release serotonin or norepinephrine but did possibly release acetylcholine.

The specific chemical nature of a neuron's neurotransmitter can be determined by histofluorescent stains. The two stains below show the blue or green glow of stained norepinephrine-containing cells in the brainstem nucleus locus coeruleus. These neurons supply the entire brain with norepinephrine via their widely branching axons.

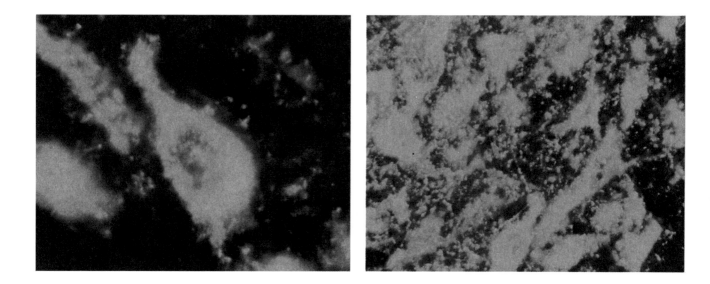

These discoveries allowed McCarley and me to formulate a chemically specific model of the REM sleep clock and trigger, illustrated in the diagram on this page. We called this hypothesis the reciprocal interaction model because of its central concept: when the aminergic (REM-off) cells were on (in waking), the cholinergic (REM-on) cells were inhibited, and vice versa. This push-pull model was chemically testable, and we have been able now to induce REM sleep by either increasing cholinergic excitation or reducing the aminergic inhibition of the REM-generating neurons. The model has a mathematical description and can be used to help explain the mechanism and nature of dreaming, as well as to provide a unified structure for comprehending the genesis of the sleep disorders. Of course, the reciprocal interaction scheme is only a working hypothesis, and other mechanisms and models may come to light as research proceeds.

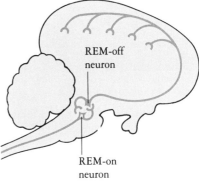

The brainstem contains two populations of neurons: one population becomes active during REM sleep (the REM-on cells), while the other becomes inactive (the REM-off cells). REM-off cells are noradrenergic and serotonergic (and inhibitory), whereas the REM-on cells are cholinergic (and excitatory). The difference in electrical and chemical properties is the basis of the reciprocal interaction model of sleep cycle control.

THE NEXT PHASE: SLEEP FUNCTION

The next phase of sleep research will explore the function of sleep. Some suppose that until the mechanistic questions of how the brain controls sleep are completely resolved, it will be impossible to know, at least at the cellular and molecular level, what sleep does for the brain. Whatever the mechanistic details prove to be, the view of sleep as simply restful is clearly inadequate. Among the likely active functions of sleep are the structural development of the brain in early life, the active maintenance of brain programs for instinctual behavior, and the active processing and storage in the brain of information acquired during waking experience.

CONCLUSION

Since sleep is a global behavior, involving the entire body, it may be looked at from many vantage points. And sleep research is appropriately interdisciplinary, embracing scientists of many persuasions and technical competencies. This introduction has attempted to show that the study of the brain is the base, the anchor, and the center of the field. Many of the ideas about sleep in this chapter are summarized in the illustration on page 24.

The biological functions of sleep are considered in Chapter 2, which discusses how the timing of sleep is an adaptive response to the environment. Chapter 3 presents adaptive functions of sleep by describing the uses of sleep by many different animals in varied kinds of habitats. That sleep may be

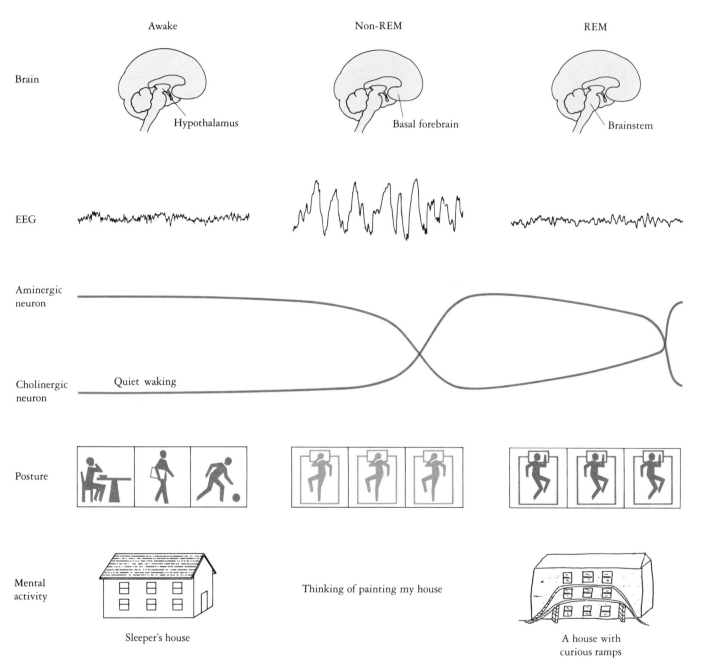

The hypothalamus, the basal forebrain, and pontine brainstem are believed to control the states of waking, REM sleep, and non-REM sleep. As we go from one state to another, a series of coordinated changes occur in EEG signals, neurotransmitter level of activity, posture, and mental activity. Posture shifts occur during the transitions to and from REM sleep. The vivid perceptions of reality in waking shift to thoughtlike, nonvisual cognition in non-REM sleep and then to the bizarre visual imagery of dreams.

creative as well as restorative is suggested by its dramatic overrepresentation in early life when the brain is developing, as shown in Chapter 4 on the development of sleep throughout life.

Our behavior does not cease with the loss of consciousness that accompanies sleep onset. Chapter 5 tracks the sequence of posture shifts and other motor acts that have been recorded in sleep and takes up the consequences of increasing, decreasing, or shifting the occurrence of sleep. In Chapter 6, the orderly sequence of brain, body, and mind events that constitute the sleep cycle is shown to depend on a neuronal clock in the brainstem, whose reciprocal chemical influences ebb and flow like fast tides throughout the night. The anatomy and physiology of sleep cycle control is thus developed in Chapter 7 to explain a new theory of dreaming, whose conclusions differ greatly from those of psychoanalysis.

The kinds of things that can go wrong with the brain clocks controlling sleep are described in Chapter 8, as a way of explaining sleep disorders. Chapter 9 concludes the book with a discussion of theories exploring the functions of sleep, theories that are now being put to experimental test. Although particular outcomes are in doubt, one general conclusion seems certain: our brains work while we sleep. And we awaken not only rested but refurbished, and perhaps even reconstructed.

2

The Rhythms of Sleep

We human beings have developed special technologies, such as air conditioning and oil burners, to protect ourselves from discomfort and death by extremes of heat or cold. What we may not realize is that our bodies have also evolved strategies for maintaining constant body temperature despite large fluctuations in the environmental temperature. One such strategy involves the unconscious brain-based rhythms that time when we sleep and wake. As this chapter will reveal, the timing of sleep is controlled by a brain clock designed for twin strategies: survival and energy conservation.

That sleep and body temperature were linked was known from the work of German physiologists in the mid-nineteenth century. They established the clinically important fact that human beings normally experience daily fluctuations in core body temperature of about 1.5°F, with a peak in the late morning or early afternoon and a trough at night—coinciding with sleep. It has since

The rise and fall of the sun causes daily fluctuations in ambient light and temperature. Organisms have adapted to these fluctuations by developing rhythms of their own. One such rhythm is the daily rise and fall in body temperature. The graph records the variations in body temperature for one human subject over the course of several days.

been well documented that this rhythmic change is not a reflex response to the more extreme daily fluctuations of ambient light and temperature. Far from it. It is, rather, a rhythm generated by the brain, which anticipates the ambient variability and fits behavior to it. Accordingly, sleep is strategically placed on the descending limb of the body temperature curve while arousal occurs when things are looking up from a thermal point of view.

Why does body temperature vary? And why do we sleep when body temperature is low? The answers lie in the connection between body temperature and metabolism. Our bodies constantly generate heat as well as obtain biochemical building blocks for our tissues, all through the set of processes known together as metabolism. To sustain this activity, we must obtain energy by consuming food.

Within the same organism, a high level of metabolic activity will coincide with a high body temperature. Because sleep is a state of lower metabolic activity, with a lower body temperature, it is energy conservative. Even if we cannot seek food in sleep, we do not use up much energy, because our muscles

Many animals, like these Barnacle geese, conserve energy by becoming immobile and assuming postures that reduce their radiant surface area. Sleep facilitates these maneuvers and is in turn facilitated by them.

become inactive. And interestingly, we do not normally become hungry while asleep. Removing individuals temporarily from the effort to find nourishment is one of the main contributions sleep makes to the survival of mammalian species. By decreasing the amount of food we need to consume, sleep makes it possible for animals to compete more successfully for limited food supplies.

The mere fact that our lives are punctuated by periods of sleep has immense benefits. Yet nature has greatly increased the benefits by carefully programming the timing of the periods of sleep. Both the alternation between sleep and waking and the related fluctuations in body temperature reoccur regularly at predictable intervals. They are thus two of the many biological rhythms that control our bodily activity. This chapter explores the timing of sleep: when it occurs, when it cannot occur, how often it occurs, and for how long it occurs.

BIOLOGICAL RHYTHMS AND BEHAVIORAL STATES

In our world, which both rotates around its axis and revolves around the sun, temperature and light fluctuate greatly each day and considerably more over the course of a year. The result is an exceeding rich dynamic of environmental conditions. The problems that nature must overcome to design animals that can successfully adapt to the constantly changing environment are, to say the least, formidable. To adapt to the variations in heat and light, organisms have developed different, though in the main parallel, strategies.

Thus, with reference to sleep and waking, there are both day creatures and night creatures, both summer creatures and winter creatures. Such temporal and thermal specializations give animals advantages over their competitors at specific times. They provide a set of betting odds as to when it would be most propitious to forage and to court—and when it would be best to repair to the nest for the energy-conserving functions of sleep.

One way that an organism could deal with the variations in heat and light is in a completely reflex fashion, that is, without the mediation of internal rhythms. When the ambient temperature rose or fell, body temperature would rise or fall accordingly. This method has two rather major disadvantages. First, the organism would be capable of only a limited range of behavior. What is worse, in such a limited mode, it would be exceedingly vulnerable to environmental extremes and would soon be either cooked or frozen. Even cold-blooded animals (which do not maintain a constant body temperature) have internal rhythms.

On a whirling globe in a solar system, time blocks such as days and seasons are reliable constants. Hence organisms were able to evolve clocks that made both their activity and their body temperature rhythmic in synchrony with the cosmic forces. And these biological rhythms soon became not only harmonious with geophysical changes but predictive of them: organisms developed a primitive kind of memory that enabled them to guess when it would be favorable to be active and when it would be more favorable to rest.

The persistence of cosmic conditions for more or less defined periods of time explains another important feature of rhythmic behavior: it is organized into a sequence of behavioral states. A biological rhythm can be considered a gradual alternation between peaks and troughs; in the sleep-wake cycle, the less energetic activity occurs in the trough and the highest level of arousal at the peak. The peaks and troughs of biological rhythms tend to be flattened so that the organism's physiological conditions can remain relatively constant for the appropriate periods. Thus, once the sun (i.e., amount of light) is above a certain level (dawn), we are likely to stay up for 8 to 16 hours (see the diagram on the next page). A smart biological clock would therefore be able to predict not only when the various environmental conditions are likely but also how long they will last, thus allowing the organism to maintain the physiological conditions most adaptive to its environment. Biological clocks also cue the animal when to honor all priorities: for example, how often and when to feed, how often and when to breed, how often and when to care for offspring. It is quite clear that such important matters could not be left to chance.

We may thus define a biological rhythm as any variable physiological process with phases that reoccur periodically. Rest/activity and sleep/waking are related but, as we shall see, they reflect different biological rhythms.

Many animal activities are governed by the interaction of the light-dark cycle with biological rhythms. Left: *The morning and evening are the most active times for this marmot, here busily munching some flowers.* Right: *Light often signals the end of sleep. Although it is still dawn, these deer are already fully awake.*

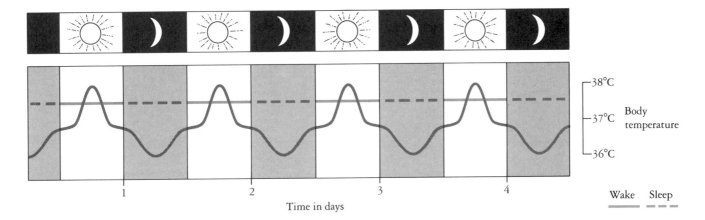

Time in days

38°C

37°C   Body
temperature

36°C

Wake   Sleep

We may define a state as any relatively invariant set of physiological conditions. Relative invariance means that both the physiological conditions and the duration of a state may vary, but only within fixed limits. The definition of state ties the concept to that of biological rhythm, in that states are phases or subphases of biological rhythms. Thus, sleep is a state that has certain defining conditions (relative immobility, recumbent or relaxed posture, and decreased sensitivity to sensory stimulation) and occurs for a finite time as a periodically recurrent phase of a biological rhythm. And REM sleep is a state within the state of sleep, as it, too, has defining conditions (brain activation, muscle tone suppression, and the REMs themselves) and recurs periodically for a finite length of time.

If the peaks and troughs of rhythms tend to be flattened into relatively constant states, then switching between states must be relatively rapid. This need is especially understandable if the states that constitute the peaks and troughs of rhythms are, as in the case of waking and REM sleep, simultaneously highly differentiated from one another and highly valuable to the survival and propagation of the species. To guarantee rapid switching, the organism will need not only to follow external cues but to have its own active ways of getting from state to state. We will see that it does.

The rhythms of body temperature (red) and sleep (orange) are controlled by brain clocks that actively anticipate the periodic availability of two forms of solar energy, heat and light. The result is that most mammals stay awake and produce heat in the daytime and they sleep and conserve heat at night.

THE CIRCADIAN RHYTHM

So advanced is evolution that even the lowliest creatures anticipate the daily fluctuations in solar energy by scheduling many activities to recur at about 24-hour intervals. Internal clocks that control biological rhythms having periods of about the length of a day are called circadian, from the Latin *circa* (about) and *dies* (day). Not only do one-celled animals and plants possess these

Left: *The single-celled alga gonyaulax emits flashes of light that are brightest at night, giving the ocean around it a blue glow.* Right: *This image-intensified video microscopic picture shows the light flashes from a single gonyaulax cell. Such "bioluminescence" arises from many tiny sources called "scintillons," whose daily activation is controlled by a circadian clock within the cell.*

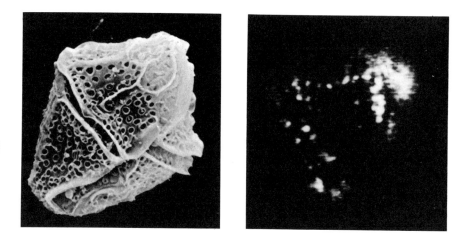

internal clocks, but it has even been claimed that DNA, the genetic material itself, is a circadian clock. And those more complicated animals that sleep as well as rest most certainly possess such timepieces. Recently, the genetic basis for circadian rhythmicity has been elucidated via the study of "clock mutants," animals with different periods of rest and activity.

Human beings normally depend on cues such as alarm clocks, sunsets, and rapid ambient temperature changes to tell us when to awake and when to sleep. Then how do scientists know that we are following an internally generated rhythm and not just reacting to these cues? The proof that circadian rhythms are internally generated, or endogenous, is provided by the persistence of a nearly 24-hour rhythm of rest/activity, sleep/wake, and body temperature when time cues are removed. Time cues are called *zeitgebers* (time givers), a German word, in honor of researchers Jurgen Aschoff and Rutger Wever, who proposed the concept.

To discover whether internal circadian rhythms actually existed, Aschoff and Wever isolated human subjects in their specially built underground bunker laboratory, which was part of the Max-Planck Institute at Erling-Andechs, near Munich. Each subject lived alone in the bunker for one month. Subjects were told that they could sleep when they wished but that they should try to consolidate their sleep into one bout per day and try not to nap. Deprived of all social cues such as clocks, the subjects did not actually know when the day began or ended; they had to rely on their own subjective estimates. When they began their daily sleep bout, they were to lie on an instrumented bed, and they were to indicate the initiation and termination of sleep bouts by turning the lights off and then on again. If they felt they had to nap, they were to so indicate by activating a special switch. The sleep and nap signals were duly charted on an event recorder. Only some of the subjects were monitored for the

EEG signs of sleep. But the core body temperature of all subjects was continuously tracked by rectal probes.

The most striking finding to emerge from the Aschoff–Wever bunker experiments, and one that has now been abundantly replicated, is that although humans, like all other organisms, have a rhythm of body temperature and sleep that approximates 24 hours, this rhythm is not identical to 24 hours. As a consequence, when either the body temperature peak or the time asleep was plotted on 24-hour time lines, the data points fell regularly and progressively more out of phase with clock time, as illustrated in the chart on this page. These rhythms thus ran at a frequency of the bodies' own devising, and because they had broken free of the entrainment normally imposed by zeitgebers, they were called free-running by their discoverers. Since most subjects had an endogenous rhythm that was longer than 24 hours, they regularly ran ahead of clock time by the difference between each subjective day and 24 hours. Thus, in 29 days of clock time, a subject who lost 51.4 minutes a day would experience only 28 subjective days.

These results mean that humans have a natural tendency to sleep at a frequency that is slightly different from what will in reality be most adaptive. The same is also true of other mammals. The circadian rhythm of sleep must thus be reset each day—in the case of most mammals, by changes in light and temperature; in our case, by alarm clocks and by work schedules, which, as we well know, are quite independent of ambient light and temperature.

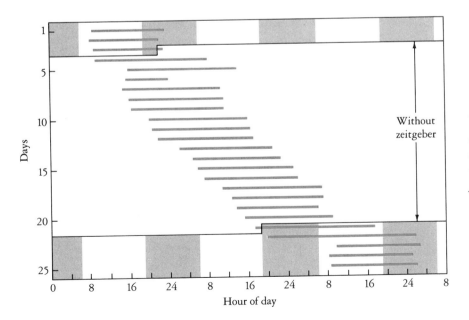

When a human subject was isolated from time cues (or zeitgebers) by living in an underground bunker, the daily activity rhythm (orange lines) changed its frequency from 24 hours (days 1–3) to 25.9 hours (days 4–21). The orange lines represent waking hours; they run in a diagonal because the subject awoke and went to sleep later every day. The rhythm frequency returned to 24 hours when the subject was again exposed to the diurnal light-dark cycle (days 22–26).

Jet Lag and How to Combat It

One of the best ways to understand circadian rhythm biology is to examine the effects of moving our own biological clocks across time zones in modern jet aircraft. The discomfort caused by the ensuing need to reset our clocks to a new local time is called jet lag. Understanding jet lag is not only educational with respect to sleep and circadian physiology, but can be of significant practical value when the principles behind jet lag are used to reduce its ill effects.

All eastbound jet travel *shortens* the traveler's day or night. Thus, as shown in the figure on this page, if you board an aircraft in Boston at 9:00 p.m. and land in Paris at 8:00 a.m., you will experience a night that is five hours shorter than the night you would have experienced had your flight been canceled. Since you will be flying "into the sun," you will experience sunrise five hours earlier than usual. You feel terrible because your body clock, whose 25-hour rhythm is reset each day at sunrise, gets only partially reset and because you have lost at least five hours of sleep. Circadian clocks readjust only gradually to such large jumps in time; in this case, it may take five days to completely reset the circadian clock.

The adjustment of body rhythm to time zone shifts is most difficult following eastbound flights, because these flights require that the time of sleep onset be advanced. It is much easier to delay the time of sleep onset than to advance it, as anyone who normally retires at, say, 10:00 p.m. can appreciate by comparing the difference between trying to go to bed at 6:00 p.m. on Friday night and staying up until 2:00 a.m. talking with friends. The best way to reduce jet lag following eastbound flights is to advance the time of awakening on the days prior to departure: will power is more effective in getting up early than in going to sleep early, and resetting is more easily accomplished gradually than all at once.

When an airplane leaves Boston at 9:00 p.m., it is already 3:00 a.m. the next day at its Paris destination. As a result of flying into earlier time zones, sunrise occurs only 3 or 4 hours later, when the passenger has just begun to sleep. On landing in Paris at 8:00 a.m. local time, it is only 3:00 a.m. in the passenger's brain, which is not only still sleepy but also sleep-deprived. The result is jet lag.

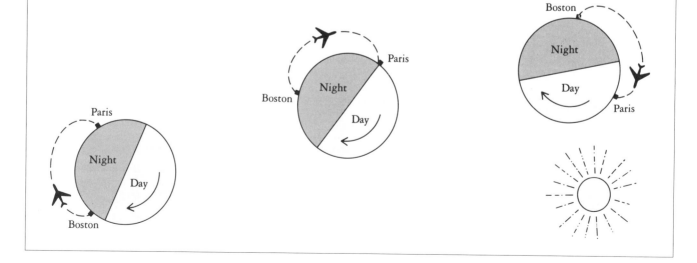

SLEEP AND BODY TEMPERATURE

The work of Jurgen Zulley and Scott Campbell, who were also affiliated with the Max-Planck Institute, adds another important dimension to our picture of sleep. Zulley and Campbell found that under the free-running conditions of the bunker a person who is feeling sleepy will usually initiate a long sleep bout only at a point near the nadir of the body temperature curve. When body temperature is falling, in contrast, sleep is virtually impossible. This means that during most long sleep bouts, body temperature is gradually rising. At all other points on the body temperature curve, including the maximum, a person may nap but sleep will not be sustained. Thus, another upshot of the Max-Planck sleep research is that the circadian rhythm in humans links sleep behavior to body temperature.

In addition to the observed correlation between sleep and the circadian body temperature curve, sleep has certain marked effects on the production of body heat. At sleep onset, body heat production is likely to fall. One reason for this is that when muscular activity is shut off, the major source of heat generation ceases. A second reason is that mechanisms that cause heat loss are affected in ways that produce cooling. The shivering response, which serves to generate body heat via activation of muscle mass, is suppressed when people fall asleep in cold environments. In warm environments, by contrast, sweating is turned on at sleep onset and promotes heat loss through the cooling associated with evaporation. The net result of these processes is that body temperature decreases as a function of sleep, as well as a function of circadian rhythm phase. And the lowered body temperature makes possible energy conservation.

Nor are these changes confined to sleep onset. When the person is asleep, heat control systems continue to devolve, to "unwind," as it were. In REM sleep, the regulation of body heat is so dramatically inhibited that, to all intents and purposes, central control of body temperature is lost and the animal relies on equilibrium with the environment—or arousal—to maintain a stable body temperature. In keeping with the findings of Zulley and Campbell, the body temperature curve has passed its nadir and is rising by the time that REM sleep is occurring in abundance during the second half of the night.

In their respective laboratories at Bologna, Italy, and Palo Alto, California, Pier Luigi Parmeggiani and Craig Heller have shown that the loss of responsiveness to changes in temperature during REM sleep is paralleled by—and thus perhaps even caused by—a loss of sensitivity to heat of the neurons of the hypothalamus, which is believed to be the brain's thermostat. To explain the function of the sensitivity loss, we are forced to go beyond the energy conservation concept. One logical speculation is that heat control neurons are rested in REM sleep (when they are not needed because the environmental temperature has already been tested and shown to be within reliable limits) so that they will be able to deal more effectively with temperature vicissitudes in the ensuing waking period.

Left: When the hummingbird temporarily enters shallow torpor, it reduces its metabolic rate to 2 percent of its normal level. Because of the low metabolic rate, it need only burn one-fiftieth the normal number of calories to stay alive. Right: In related energy conservation strategies, some animals go into deeper hypometabolic states and stay there for long periods of time. The marmot is hibernating in his burrow in winter, when the metabolic cost of maintaining normal body temperature greatly exceeds the availability of calories from the food supply.

OTHER DORMANT STATES

Because of the central importance of heat regulation and sleep in mammalian life, it is of interest to examine two specialized heat conservation states, shallow torpor and hibernation, for insights about sleep.

Shallow torpor is a state of dormancy found in small mammals and birds, in which profound drops in body temperature occur on a daily basis in conjunction with inactivity and sleep. Experiments are difficult in such small creatures as the pygmy mouse and the hummingbird, but temperature drops from 98.6°F to between 47.1 and 32.6°F have been recorded. Occasionally, a bout of torpor may extend over several days, in which case emergence from dormancy occurs at the expected time of arousal from ordinary sleep, indicating the persistence of circadian control. When these extended bouts occur

during the hot, dry summer months, the phenomenon is referred to as estivation. As in the slow wave sleep of other mammals, the temperature drop of shallow torpor is not a passive response to ambient temperature but is regulated by the brain. This indicates that active brain mechanisms are at work in orchestrating adaptive thermal strategies and reinforces the view that sleep, too, is actively generated and regulated.

Deep torpor, or hibernation, is a more prolonged, seasonal drop in body temperature to 0°C (32°F). It is an adaptation found only in a few species of less than one-third of the orders of mammals. Like shallow torpor, it is seen only in small mammals, the largest being the marmot, which weighs between 6 and 10 pounds. Small animals have relatively large surface-to-volume ratios, which makes them, like human infants, more vulnerable to radiant heat loss. Hibernation may thus be a way for these animals to avoid expensive and possibly fatal exposure to cold. Since they cannot efficiently maintain a high body temperature, they "drop out" and "cool it" in the safety of hypothermia.

Studies have shown that animals usually enter into hibernation through what appears to be physiological sleep: in the first part of the temperature fall, from 37 to 25°C, either non-REM and REM sleep occurred in their normal proportions (80 percent and 20 percent, respectively) or there was a progressive decrease (to 10 percent) of REM sleep. Below 25°C, recordable electrical activity ceased, and these stages could no longer be discriminated. Nonetheless, even during deep hibernation, there was other evidence of periodic brain activation, indicating that the brain continues to regulate dormancy even in physiology's coldest known icebox. Since animals "know" when to emerge from hibernation—just as we "know" when to wake up—this should not be surprising.

According to Craig Heller, sleep, shallow torpor, and hibernation are three comparable forms of dormancy, whose low metabolic cost represents the mammalian response to selective pressures favoring energy conservation.

A CIRCADIAN CLOCK IN THE HYPOTHALAMUS

Since the mammalian sleep cycle is nested within the trough of the circadian rhythm, we would like to know how the intricate system controlling circadian rhythm works. The early work of Curt Richter at Johns Hopkins University set the stage for the more detailed studies of the modern era. By recording the activity of rats on running wheels, Richter was able to locate the probable site of the circadian clock in the brain. Rats are usually active on the running wheel only during the night. Richter found that he could make the rhythm of activity free-running by blinding the animal, which deprived it of

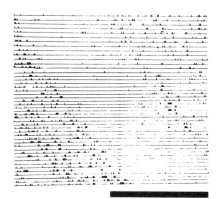

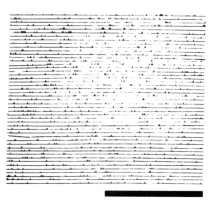

The organization of normal daily activity rhythms is a function of the circadian clock in the hypothalamus, as was demonstrated by these experiments of Robert Moore on rats. The chart at the left plots the daily water-drinking activity of a normal rat. When blinding deprived the brain's circadian clock of the light that served as its time cue (middle), the activity rhythm became free-running at a period different from 24 hours. And the rhythm became completely disorganized (right) when the hypothalamus was damaged. The bar at the bottom indicates the dark part of the light-dark cycle.

the light cues of daylight. More interesting, he found that all activity was eliminated by damage to the hypothalamus, a deep, central brain structure that has abundant interconnections with other brain regions. The implication was that the hypothalamus housed a circadian clock that beat at a frequency of once every 24 ± 0.5 hours and that this circadian rhythm was entrained to its pattern of daylight activity by sensory input, especially light. The results of an experiment similar to Richter's are summarized in the diagram on this page.

Wishing to further pinpoint and analyze this clock, Robert Y. Moore, then at the University of Chicago, made smaller experimental destructions and found that the crucial region was the suprachiasmatic nucleus (SCN), so named because each of its paired parts sits in the hypothalamus just above the optic chiasm, the point at which the nerves from the two eyes cross on their way to opposite sides of the brain. And this strategic location is not coincidence, as Moore demonstrated when he found a direct projection of primary visual fibers from the eyes, via the chiasm, to the SCN. This retinohypothalamic tract, as it is called, thus provided a pathway for the entraining effects of light upon the circadian clock.

The fact that the hypothalamus is a small structure with many specialized cell groups, each of which contains very small neurons, has made it difficult to study this critical center by recording the activity of single neurons. Thus we cannot at present specify the link between the suprachiasmatic nucleus and the more easily studied systems of the lower brainstem that control the sleep cycle. That such a link must exist is clear, both because normal sleep follows a circadian rhythm and because circadian control is disrupted by transections of the midbrain, which separate the hypothalamus from the pons, or by obliteration of the hypothalamus.

We also know, however, that the circadian rhythm of the SCN is reflected in the firing pattern of the SCN's own neurons. When the suprachiasmatic nucleus is surgically isolated or when the brain tissue containing the SCN is removed and kept alive in organ culture, the long-term recording of multiple cells in either preparation reveals that their firing waxes and wanes according to a circadian cycle. These studies, carried out by Shin Ichi Inoue and Hiroshi Kawamura of the Mitsubishi Institute in Tokyo, show that mammalian brain-cell aggregates can keep circadian time and translate that capacity into electrical signals. Those signals might be used by other parts of the brain to synchronize their own activity with the energy and information fluxes of the outside world.

But what about the chemistry of this clock and the bells it rings each day throughout our brains? Although the hypothalamus is the brain's hormonal factory, we do not yet know if one or more of its many molecular products is a sleep signal. The suprachiasmatic nucleus alone contains several long-chain proteins (or peptides, such as vasopressin, somatostatin, and neurophysin), all of which are logically—but as yet not experimentally—connected to sleep.

In summary, the suprachiasmatic nucleus of the hypothalamus appears to contain a circadian clock whose output is essential to synchronize the activity of the rest of the brain—and hence the body—in such a way as to bring it into

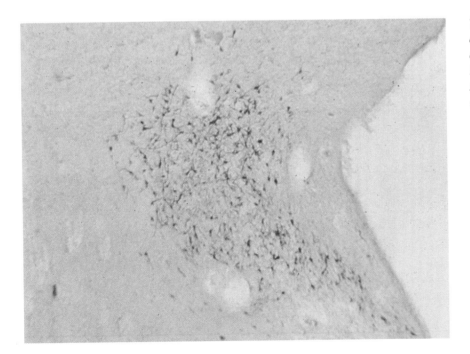

The human suprachiasmatic nucleus is clearly visible as the darker area in the center of this photomicrographic image by Robert Moore. The neurons have been stained to show the presence of the hormone vasopressin.

daily harmony with the cycles of heat and light. Although the precise neural mechanisms are not known, it is clear that sleep is a key energy-conserving state that is under circadian control. I now turn my attention to the organization of sleep per se.

THE REST/ACTIVITY CYCLE: AN ULTRADIAN RHYTHM

Long before the discovery of the non-REM/REM cycle of sleep and dreaming, the Chicago physiologist Nathaniel Kleitman organized his observations on the fluctuating levels of arousal and motor behavior during the *waking* period of animals (including humans) into his concept of a basic rest/activity cycle (or BRAC). Kleitman thought that in addition to the circadian rhythm there might be another rhythm, which ran at a shorter period. A key source of data supporting this concept was his study of newborn human infant behavior, which will be discussed in Chapter 4. But a moment's self-reflection will help us all understand the importance of Kleitman's idea: our waking lives are subdivided into temporal packets. To see this, we need only consider such obvious time blocks as school class periods, coffee breaks during work or "happy hours" after work, or even just our need to get up and walk or talk— or to fall silent—after not having done so for some time. Sit still for a bit and something will move you. Go for a while and you will naturally seek rest. The illustration at the bottom of the page shows how the day could be broken into BRAC periods.

Kleitman's own discoveries, made in conjunction with his co-worker Eugene Aserinsky, proved him to be correct—at least as far as sleep is concerned. Because sleep turns out to have an internal rhythm with a period length of 90 to 100 minutes, and because the frequency of this rhythm is

This schematic rendition of a human day divides the 24-hour span into 16 segments to illustrate Kleitman's concept of a 90-minute basic rest-activity cycle (or BRAC). The first five boxes are the subdivisions of a 7½-hour sleep period by the non-REM/REM cycle. The silhouette figures in the following boxes represent one way that waking activities could be organized in BRAC periods. The great flexibility and variability of our daily schedules indicate that the brain clocks controlling the timing of waking behaviors can adapt to widely varying environmental conditions.

12 midnight   6 a.m.   12 noon   6 p.m.   12 midnight

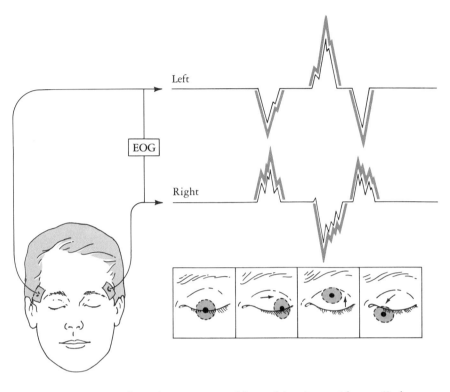

The rapid eye movements of sleep are detected in sleep labs by recording the potentials generated when the two retinas move toward or away from electrodes placed outside each eye. Sleep lab technicians normally record the movement of each eye separately in such a way that the signals move in opposite directions on the paper record. This enables them to distinguish eye movement from the EEG potentials of the anterior pole of the brain, which are also picked up by the eye electrodes.

greater than one per day (if uninterrupted by waking it would actually be very close to 16 per day), we call it an ultradian rhythm.

THE DISCOVERY OF REM SLEEP

As mentioned in Chapter 1, the discovery of REM sleep was made possible by Aserinsky's decision to record eye movements in children. Aserinsky at first focused his studies of attention on children who were awake. He had noticed that eye movement correlated with changes in alertness within the waking state, which the EEG was too insensitive to detect. The less alert the child, the less his eyes moved. Think for a moment of what happens to the eyes of someone who has lost interest in what you are telling him: they glaze and stare fixedly, indicating a withdrawal of attention from the outside world to an inner world, perhaps of fantasy, perhaps of sleep.

Aserinsky recorded eye movement by attaching EEG electrodes to the skin near the eyes; the electrodes picked up the voltage change that occurred when the two retinas moved toward (or away from) his electrodes (see the diagram on this page). He used this electrooculogram, or EOG, together with the EEG to define sleep onset more precisely.

When the children closed their eyes and fell asleep, Aserinsky to his surprise observed that there were clusters of eye movements. These eye movements occurred at the same time as the EEG of Stage I. With the encouragement of his mentor, Nathaniel Kleitman, Aserinsky quickly shifted his attention from waking to sleep. On recording the brain waves and eye movements of adult subjects, Aserinsky and Kleitman found that every 90 to 100 minutes periods of EEG activation and eye movement lasting from a few to many minutes followed periods of EEG synchronization and eye stillness. Because these "emergent" Stage I EEG epochs were associated with rapid eye movements, they were called Stage I REM in contrast to the "descending" Stage I epoch at sleep onset, which did not have the REMs. Aserinsky and Kleitman also found that this EEG cycle involved not only the brain, but also the heart and the lungs as is shown on page 44. Specifically, cardiac and respiratory rates both increased during the Stage I–REM epochs—a clear sign that sleep could no longer be regarded as a state of quiescence for these vital functions. And most interesting of all were the results obtained by another of Kleitman's students, William Dement. When he awakened subjects from Stage I–REM sleep, they were able to provide detailed reports of dreaming (see Chapter 7).

As Pasteur said, "In the field of observation, chance favors the prepared mind," and the discovery of REM sleep is a case in point. The initial observation was, of course, accidental, and it was also lucky in that with child subjects, it was possible to observe REM sleep at sleep onset, whereas with adults REM sleep generally first occurs only after 90 minutes. Equally important, however, Kleitman's mind was well prepared to recognize the likelihood both that the rapid eye movement in sleep would prove to be associated with dreaming and that it represented a periodic activation of the brain in sleep akin to his BRAC.

THE NON-REM/REM CYCLE

By the mid-1960s scientific research had produced several valid generalizations regarding the EEG sleep cycle. Based on the recordings of hundreds of male and female subjects, it was found that the cycle length—the time from onset of sleep to the end of the first REM period or from the end of the first REM period to the end of the second REM period and so on—was relatively constant for any given subject and for subjects of the same age (see Chapter 4). The proportion of each cycle devoted to Stage I REM was, however, variable. This proportion was usually quite small in the first cycle but quite large in the third and fourth, so that one-half of the REM sleep occurred in the last third of the night. Non-REM sleep had, of course, the opposite distribution, occur-

Nathaniel Kleitman and one of his canine collaborators in physiological studies at the University of Chicago in 1925. With his students Eugene Aserinsky and William Dement, Kleitman later revolutionized sleep physiology and psychology with their 1953 discovery of REM sleep and its relationship to dreaming.

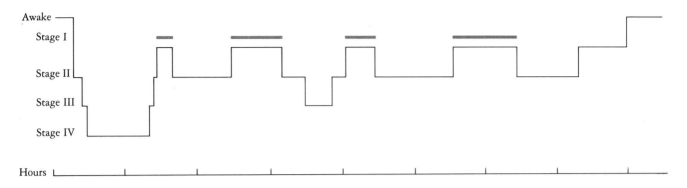

ring more in the first and second cycles than in the third and fourth. This unequal distribution was particularly characteristic of Stage IV, the most synchronized phase of the non-REM sleep, two-thirds of which occurred in the first third of the night. Like cycle length, the percentage of total sleep time spent in Stage I REM, or in Stage IV, remained relatively constant from night to night for any given subject. The graph on this page shows the changing length of sleep stages in one person's sleep.

Examination of the sleep cycle graphs describing this pattern suggested several intriguing hypotheses. The constancy of the length of each successive non-REM/REM cycle suggested that the cycle was controlled by a reliable timer somewhere in the brain. As pointed out in Chapter 1 and detailed in Chapter 6, this clock turned out to be located in the pons. The fact that the earlier cycles were deeper (having more non-REM sleep) and the later ones shallower (having more REM sleep) suggested that the amplitude of the non-REM/REM cycle was dependent on some other process—such as the circadian rhythm. The superabundance of Stage IV sleep early in the night suggested to others that non-REM sleep was a response to fatigue or to the duration of the preceding period of wakefulness or to both.

This rest-and-recovery hypothesis fits well with the data regarding brain and body temperature in supporting the view that non-REM sleep, which makes up 75 percent of total sleep time, is a hypometabolic state. That is, the metabolic cost of non-REM sleep is low, and the body expends little energy. Also consonant with this view were the low levels to which heart and respiratory rates fell during Stage IV, the trough of the cycle. That brain activity was at a low ebb was indicated by experimenters' difficulties in waking subjects up from Stage IV and by the confused, garbled reports they elicited when they attempted to interview these groggy subjects. Anyone who has tried to get up and going in response to an alarm clock or phone call in the first three hours of sleep knows this feeling well. It may take a full 15 minutes of active mobilization to achieve even a semblance of mental clarity and well-being!

After recording up to 1000 pages of EEG record on a single night, sleep lab technicians score each page to determine the minute-by-minute progression of sleep states. Then they plot the data in the graphic form shown here. This particular graph shows that the subject's sleep was uninterrupted and that it consisted of four non-REM/REM cycles. In the first cycle, non-REM is deeper and longer than later in the night, when the REM periods increase in length (purple bars). In a later, fifth cycle, just before awakening, there was a long Stage I EEG epoch without REMs. Many of the graphic features shown here change with age and with disease, making EEG recordings an indispensable tool in documenting normal and abnormal sleep.

A closer look at one non-REM/REM cycle reveals other dynamic sleep features that can be captured by the polygraph. The eye movements of Stage I–REM sleep occur in clusters, which are longest and most intense early in the epoch. Blood-pressure level (mmHg), respiratory rate (c/30s), and heart rate (c/30s) are higher in REM than in non-REM sleep, as is their minute to minute instability. The major body movements tend to occur just before and just after the Stage I–REM period in transition from—and back to—non-REM sleep.

Later in the night, the cycle amplitude decreases. In the last two cycles, there may be neither Stage IV nor Stage III. Thus when the REM sleep peaks are compared to the non-REM sleep troughs, the difference is of only two stages, not four. And the related differences in heart rate, respiratory rate, and mental activity also become less pronounced. It becomes much easier to awaken subjects during non-REM sleep, and they more rapidly achieve a state of coherent thought and speech.

I will examine the issue of dreaming in relation to the non-REM/REM cycle in more detail in Chapter 7. For the present, it suffices to say that at all times of night, the tendency for subjects to report mental activity characterized by visual imagery, convinced and uncritical involvement in an ongoing scenario, odd or impossible happenings, and strong emotion is greatest in

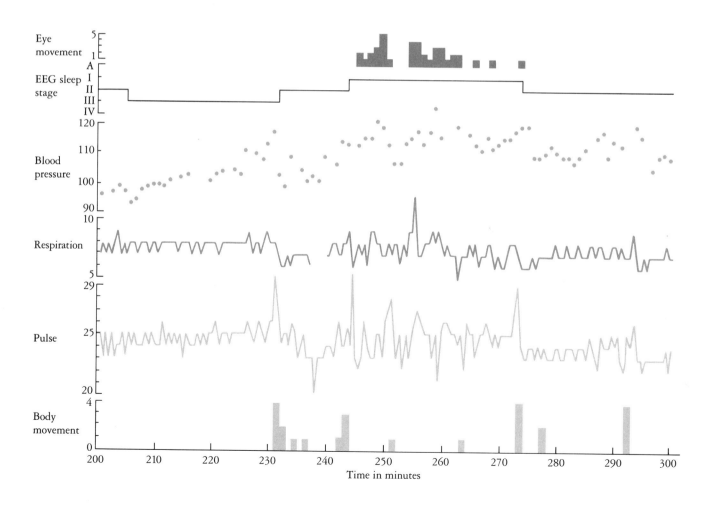

REM sleep. Reports following REM sleep awakenings are consistently longer than those given after non-REM sleep awakenings. If there is any recall from non-REM awakenings, it tends to be more thoughtlike, often involving repetition of ideas that do not progress.

That the non-REM/REM cycle affects many bodily functions can be appreciated by consulting the figure on page 44, which shows the orchestration of motor, autonomic, and brain functions that make sleep physiologically symphonic.

THE MORNING AFTER

Is Kleitman's BRAC still there? After seeing his theory emphatically reinforced by the non-REM/REM cycle during sleep, we may find it anticlimatic to return to the world of waking, but such a return will help summarize this chapter and point the way to some related topics of interest.

It is clear that there are two very robust rhythms of human physiology—the circadian rest-activity cycle (about 24 hours) and the ultradian sleep-dream cycle (about 1.5 hours)—and that they powerfully interact such that the amplitude of the latter is maximal at or near the low point of the former. As arousal increases, there occurs a powerful damping of the amplitude of the ultradian cycle. It is probably this powerful damping that has made it so difficult to decide whether or not our clearly segmented waking periods are determined by the same internal clock that determines the sleep-dream cycle.

There are some good reasons to keep open the question of whether BRAC exists during waking and to try harder to find a definitive answer. Among them is the fact that the probability of seeing REM sleep in daytime naps clearly depends on when the nap occurs in the circadian cycle. This means that the threshold of the REM sleep generator may vary continuously around the clock and not just in sleep. Another reason is the fascinating but complicated relationship of circadian rhythm phase, time of REM sleep onset, and daytime mood in patients with manic-depressive disease. This data suggests that at least in these patients, the form and the quality of sleep vary significantly as a function of when in the circadian cycle sleep is entered. The only way to explain this data is again by a model of continuous interaction between the circadian clock in the hypothalamus and the timer of the non-REM/REM cycle in the brainstem.

Add to these the results of studies showing that human subjects eat, ambulate, and cogitate at a period of 90 minutes and we see good reason to view sleep and waking as not only complementary but also continuous in the sense that both may be under the active control of the same brain mechanism.

C H A P T E R

3

Sleep and Evolution

Why is it that non-REM sleep is so much older in evolutionary terms than REM sleep? Why do lions sleep at least two times as much as their prey, the antelope, although both species are of roughly the same evolutionary age? In general, small animals sleep more than large animals, so that size and amount of sleep are related. Why, then, does one small animal, the bat, sleep ten times as much as another animal of roughly the same size, the shrew? Species differ significantly in both hours of sleep and longevity. Can a long life be in part a factor of long sleep?

By looking at how evolutionary age, brain maturation, body size, and such ecological factors as prey-predator status interact to determine sleeping behavior, this chapter provides answers to these and other questions. In general, the data reinforce the three major themes of this book.

Shakespeare once called sleep "great nature's second course, chief nourisher in life's feast." Echoing the poet, this spotted leopard sleeps in a tree with its unfinished gazelle lunch safely stashed on the branch behind him. Mathematical models of reciprocally fluctuating levels of prey and predator populations, as in the curves above, have been used by neurophysiologists to describe the changing levels of activity in brainstem neuronal populations during the sleep cycle of domestic cats.

That sleep is *of the brain* is made clear by the strong correlation between the degree of brain complexity and the complexity of sleep. Brain complexity and sleep complexity are related because the brain is the structural basis of sleep.

The principle that sleep is *by the brain* is generally supported, but the nature of the particular brain structures involved is paradoxical. REM sleep is clearly generated by the brainstem, yet many low-order animals with well-developed brainstems have none of it. The reason is that only when the fore-brain and cortex are highly evolved does the brainstem come into play. There must be an interaction between newly evolved brain structures and older ones, with the emergence of properties not present in either alone.

The principle that sleep is *for the brain* has already been qualified some-what by the metabolic considerations of Chapter 2: obviously, it is the whole organism, and not just the brain, that benefits from reducing metabolic cost. The ecological data presented here strongly underline the point by showing how certain behavioral adaptations allow animals to sleep a great deal and thereby increase metabolic efficiency. Moreover, these data also show that if ecological factors make it *disadvantageous* to sleep, some animals can get by with very little. And yet, a closer look reveals that it is not only metabolic efficiency, but also cognitive versatility, that seems to correlate with the most highly evolved sleep behavior. Might sleep have a role in information processing?

WHY DID SLEEP EVOLVE?

The survival of a species depends on individuals competing successfully for energy supplies, reproducing themselves, and protecting offspring until the offspring can themselves successfully compete and reproduce. At first glance, sleep seems to be the absence of these very goal-directed behaviors. How then, could sleep serve the evolution of mammalian species, including man?

Two answers come quickly to mind. One is direct: sleep is energy conservative. The other is indirect: sleep behavior, which in one sense is the height of vulnerability, as both sensory awareness and motor responsiveness are lost, is at the same time elaborately protective. Animals seek, build, and protect sleep sites with great diligence and care. And thus offspring, whose sleep demands during early development are twice as great as their parents', are protected. Moreover, sleep behaviors developed for protection may reinforce useful waking behaviors. For example, pair bonding, a critical factor in pro-

moting survival of offspring, is favored by nesting and by the proximity of breeding pairs and offspring that is part of efficient sleep behavior.

A more penetrating and subtle view of how sleep may favor survival and propagation begins with the recognition of the importance of information to the survival of the individual. That a proper ordering of information in the brain favors survival and propagation seems obvious. Consider the following questions: (1) How are instinctual programs such as feeding, fighting, and fleeing stored in the brain and then, when needed, activated? (2) How does an animal keep track of his orientational data? How does he know who are his suitable mates and family? How does he know what territory is home? How

Huddled warmly together in their nest, the infants belonging to this litter of dusky antechinus, a shrewlike marsupial, have slept in safety during most of their first eleven weeks of life while the brain readies itself for the risky adventures of adult life.

does he order time so that behaviors occur in appropriate and efficient sequence? (3) Supposing he has a language, consciousness, feeling, and abstract thought, how does he keep his vocabulary and grammar, ideas, and emotional repertoire up to date when he might not use a word, an idea, or an emotional expression for days, weeks, or even years?

My answer to these questions, explored in more detail in Chapter 9, can be stated simply here: information stored in the brain is actively renewed and reorganized. And while this could be done, and no doubt is done, during waking, it could also be done, and probably is done, during sleep. We know it is possible that the organism could be "learning" in some way during sleep since the brain is continuously active in sleep and becomes especially so during REM sleep. As you will see in Chapters 6 and 7, this brain activity is so highly organized that the conclusion that information is being processed is inescapable.

Waking and sleep could contribute to the renewal of information in significantly different ways. Two facts support such a view. One is that during sleep the organism, being off-line as it were, is presented with no new information from the environment to process. This means that both the software (the activation programs) and the hardware (the neurons activated) that are used could be quantitatively and perhaps even qualitatively different from those used during waking. The other is, of course, the fact that there *are* two quite different states of brain activation: one is waking, and the other is REM sleep.

I propose that there are two different but complementary modes of information processing. During waking we acquire information and apply it; we both learn and act. During REM sleep, we reinforce our learning of significant information, but we do not act. Instead, we recall and we imaginatively enact in dreams. My point is that these fictive acts may be quite useful to the brain in and of themselves in the same way that visualizing a perfect performance helps a gymnast or diver prepare for competition. These speculations about the function of sleep find support from an examination of how sleep has actually evolved as new species have developed.

THE EVOLUTIONARY PICTURE

About 220 million years ago, the advanced reptiles were the most abundant land animals of the vertebrate type. (Invertebrates were—and, of course, still are—considerably more numerous, in terms of both species and individuals.)

These reptiles had reached a level of evolution very near to that of true mammals, which appeared on the scene about 180 million years ago and may have resembled shrews in appearance and behavior. The study of extant crocodilian reptiles supports the idea that this evolutionary stage was crucial in developing the brain and behavioral patterns that scientists now study as sleep. The chart

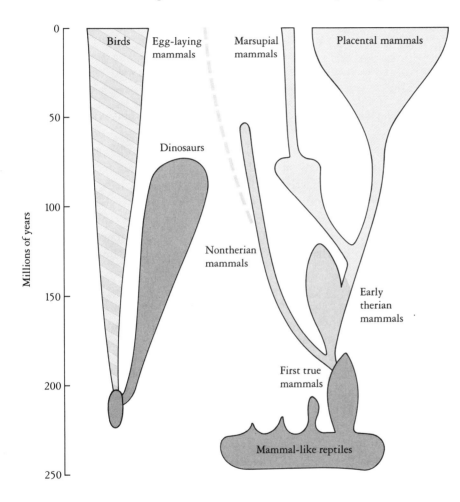

Birds

Egg-laying mammals

Marsupial mammals

Placental mammals

Dinosaurs

Millions of years

Nontherian mammals

Early therian mammals

First true mammals

Mammal-like reptiles

REM sleep

REM sleep only in hatchlings

Non-REM sleep

Unknown

The history of sleep is shown in these family trees of our animal forebears. Two lineages, each over 200 million years old, have led in separate paths to today's variety of sleep styles. With other placental mammals we humans share the alternation of non-REM and REM sleep phases. With the birds we share the preponderance of REM sleep at birth, which is surprising in view of our different origins.

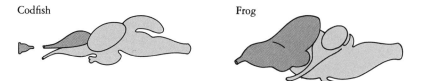

Codfish   Frog

The brain has grown in size and complexity during evolution, mainly owing to the expansion of the cerebrum. As more and more cortex has been added, giving the animal more and more behavioral versatility, the surface of the brain has become increasingly furrowed. Since sleep, especially REM sleep, has increased in parallel with cortical growth, it is natural to suppose a functional connection between sleep and cortical complexity, even though the REM state is controlled by the older brainstem.

on page 51 shows how sleep has developed with the evolution of the vertebrates.

Two groups of mammals sprang from the first, shrewlike mammal. The nontherian mammals, most of which became extinct, were egg layers like the reptiles but had the distinctly mammalian features of body hair, body temperature regulation, and nursing of the young. The two surviving species of nontherians are the platypus, which is difficult to study in captivity, and the echidna, which has been studied under laboratory conditions. As discussed later in the chapter, the echidna has well-developed non-REM but no REM sleep whatsoever.

The second main group, the therian mammals, gave rise about 130 million years ago to the two major classes of living mammals: the marsupials, like the kangaroo and opossum, which tote their immature young around in sacs, on backs, or in pockets, and the placentals, like us and the other furry animals, which have a longer time in utero. All marsupials and placentals studied to date have both non-REM and REM sleep, although with fascinating variations in timing and amount.

A separate group of creatures, the thecodonts, evolved probably from the reptiles. From this source sprang not only the dinosaurs, about whose sleep we can know nothing, but also the birds, whose sleep is easily studied. Interestingly, the birds have evolved sleep patterns that are protomammalian, showing non-REM sleep at all ages but having robust REM sleep only in early infancy.

From the perspective of sleep patterns, each vertebrate species falls into one of four broad categories: creatures that rest but do not sleep (e.g., fish and amphibia), those that have only non-REM sleep (e.g., the lower reptiles), those that have non-REM sleep and very partial, temporary, or brief REM sleep (e.g., the higher reptiles and birds), and those that have the fully developed bicyclic picture (the mammals).

Alligator   Goose   Horse

As vertebrate species become more advanced, the major change in the brain is the growth of the cerebral hemispheres. (See the figure on these two pages.) Creatures in all four categories have well-developed brainstem structures, but a true cortex is present only in the mammals. Within the different species of mammal, the cortex is more elaborate in the primates, especially homo sapiens. The developed cortex makes possible more complex behaviors, up to and including language and thought. Significantly, the animals with a developed cortex are the animals with REM sleep. Despite the huge variability in REM sleep among species that I shall shortly consider, there appears to be some causal link between REM sleep and cognitive ability. The existence of such a link would be one more support for the theory that information processing occurs during REM sleep.

WHICH ANIMALS DO NOT SLEEP?

Sleep is distinguished from rest by decreased sensitivity to sensory stimulation, from torpor by the maintenance of body temperature, and from coma by ready reversibility. And scientists now further restrict the definition of sleep to those states that both fulfill all of the above criteria and are accompanied by specific electrographic signs.

Before the discovery of the EEG, there was no way for a naturalist to distinguish between the immobile, unresponsive behavior of one animal and another. Thus, the nineteenth- and early-twentieth-century literature contains reports of sleep behavior extending all the way down to the unicellular protozoans. In fact, the protozoan infusorian, the metazoan littoral convoluta, the turbellarian worm and planaria, the crustacean crab and lobster, the molluscan octopus and squid, the insect housefly and butterfly all show diurnal or more

This dew-dropped dragonfly has awaited the dawn immobile on its perch. Its inert state is called dormancy to distinguish that state from sleep. Because insects lack temperature control, they simply become inactive in the cold.

frequent alternations of activity and rest. But to equate rest and sleep is to confound them. I will therefore assert that sleep does not exist in these animal phyla.

When it comes to the fish, it is less easy to be sure, as fish assume immobile postures. But the decreased sensitivity to sensory stimulation and the distinctive EEG patterns of brain sleep are not present, as shown by the EEG patterns on the next page. Similarly, among the amphibians, neither bullfrogs, which remain constantly alert, nor tree frogs, which are intermittently unreactive, have any electrical signs of brain sleep.

Active

EEG

EMG

At rest

EEG

EMG

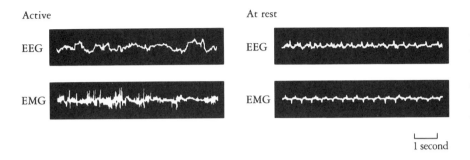

⊢—⊣ 1 second

The French perch (or tanche) shows no dramatic change in brain wave (EEG) activity when it becomes inactive (as indicated by the decrease in muscle potentials or EMG). Like the codfish cerebrum depicted in the illustration on page 52, the cerebrum of the perch is as small as its functional state is undifferentiated. Its state of rest is thus not the same as ours when we sleep.

SLEEP IN REPTILES

That sleep behavior may begin with adaptation to life on land is suggested by the laboratory findings for the various reptiles, all of which have non-REM and some of which may even have REM sleep. Four aspects of terrestrial adaptation immediately spring to mind as potentially relevant to the evolution of sleep: the need to adapt to thermal instability, as air temperature is less stable than water temperature; the need to find salt and water in order to maintain the proper salt and water balance; the need to breathe air in order to obtain oxygen; and the need of terrestrial animals to support their weight against gravity, since air is less buoyant than water. So that these needs can be met, considerable elaboration of brain control mechanisms is necessary, and in each case, the mechanism elaborated is located in the brainstem, from the

One way to survive the perils of terrestrial life is to carry your house on your back. When it emerges from the sea, this green sea turtle protects its soft parts in a hard carapace in which it can also, safely, sleep. And although it is incapable of regulating its body temperature, the turtle's brain is elaborate enough to show some of the electrical changes of human sleep.

In addition to listening to the dream reports of his patients, the late New York psychoanalyst Edward Tauber observed and recorded the sleep of such animals as this iguana lizard, asleep on a perch in Tauber's Westchester living room. Despite its huge eyes and large visual brain, this species does not have REM sleep.

medulla (respiration) through the pons (posture) to the hypothalamus (thermoregulation and hormonal control of salt and water balance). In each case, sleep temporarily reduces the demands of these regulatory tasks, and in each case, the neurons controlling these processes change their output in sleep.

Chameleons, lizards, turtles, and caimans have all been studied under laboratory conditions, and all show clear-cut brain wave changes with the adoption of characteristic rest postures, in which muscles relax and eyes close. As shown by the recordings in the margin, the turtle EEG shifts from a low-voltage, fast (11 to 13 cycles per second) pattern to a higher-voltage, more irregular, and slower (6 to 8 cycles per second) pattern. The changes in brain pattern frequency in the chameleon are similar. It seems generally to be the case that in reptilian sleep the EEG frequency drops by about half while the voltage doubles. These robust changes are comparable to those in mammalian sleep. Their magnitude indicates major shifts in functional brain organization.

The late American psychiatrist and naturalist Edward Tauber reported that eye movements occurred in clusters every 12 seconds or so during chameleon sleep but without any sign of activation in the EEG. The French physiologist Jacques Peyrethon found no such eye movements in the EEG-defined sleep of either iguana lizards or pythons. To his surprise, however, the caiman, a nonadvanced reptile, showed not only REM clusters but definite if brief periods of brain activation. Closer study defined these crocodilian "remlets" as lasting on average 50 seconds and found that they were accompanied by fine movements of the digits of the front paws.

The ambiguous sleep signs in all these species are probably best taken as evidence for the progressive evolution of mechanisms that will ultimately become the unequivocally defined sleep-generating systems of the mammalian brain. Thus, some of these animals representing the transition from aquatic to terrestrial life have none, others have some, and yet others have—evanescently—all of the signs of mammalian sleep. This same evolutionary principle is demonstrated in another way in the birds.

The turtle shows unequivocal signs of non-REM sleep: as its muscle tone falls, the turtle's EEG clearly slows and the EEG voltage clearly increases.

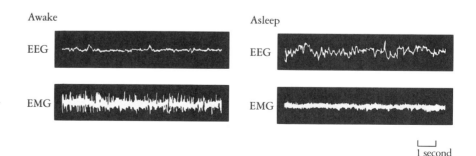

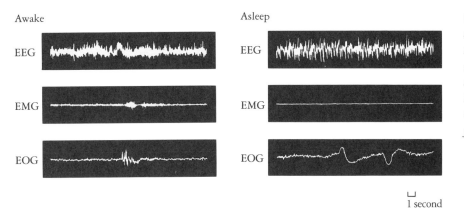

Awake

EEG

EMG

EOG

Asleep

EEG

EMG

EOG

⊔ 1 second

Unlike the turtle, the caiman displays an occasional eye movement during sleep, leading some scientists to conclude that these crocodilian reptiles (like the chameleon) are capable of brief periods of REM sleep. As explained on page 41, movement of the eyes causes voltage fluctuations in the EOG.

SLEEP ON THE WING?

The first laboratory study of avian sleep was conducted by the French physiologist Marcel Klein, who recorded the EEGs of three leghorn chickens. Surprisingly, it appears that by a completely different evolutionary path, the birds have evolved a sleep system with all the features of mammalian sleep. Yet these features are organized in a quantitatively different way, in that REM sleep is concentrated in—indeed, virtually restricted to— the period just after birth. Nonetheless, this concentration parallels the observed abundance of REM sleep in newborn mammals. The early abundance of REM sleep across

Adult birds have given up the REM sleep they enjoyed as hatchlings but are otherwise quite talented sleepers. Using their own downy wings as pillows, this flock of sanderlings demonstrates the trick of sleeping while standing one-footed in the morning sun.

Awake

EEG

EMG

EOG

Non-REM sleep

EEG

EMG

EOG

REM sleep

EEG

EMG

EOG

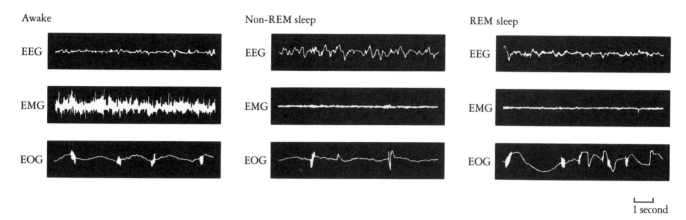

1 second

Laboratory recordings reveal that a chick loses muscle tone (EMG) and twitches its eyes (EOG) at the same time that its EEG changes to an activated pattern resembling that seen in waking and clearly different from the higher-voltage, lower-frequency EEG of non-REM sleep. The association of REM sleep with early development in birds suggests that this state may play an active role in building the brain.

species supports the idea that REM sleep plays an important role in development, an idea that will be discussed in Chapter 4.

Researchers have found ample evidence of non-REM sleep in birds. As in the reptile species discussed, birds were immobile and unresponsive for defined periods, during which periods their EEGs slowed. REM sleep, however, was not only more evident in birds than in reptiles, but it also included the loss of muscle tone seen in mammals. In chicks, REM sleep has been reported to occupy as much as 7.3 percent of each recording period. In adult birds, REM sleep appears to occupy less than 0.3 percent of the time spent asleep. The rarity of this behavior in adults is striking.

In seeking to interpret such data, Klein emphasized an ecological factor that we now know to be quite important: the fact that the chicken is a prey species makes it quite undesirable for chickens to be unresponsive *and* without muscle tone. In this view, the young must be seen as either protected, expendable, or unpalatable; but their indulgence in REM sleep also betrays a probable developmental advantage that is taken in trade for increased vulnerability.

To test Klein's hypothesis, Tauber and the Mexican physiologist José Rajas-Ramirez decided to record from adult members of a predator avian species. They found that the sleep of four falcons was characterized by unresponsiveness and immobility, with occasional EEG slow waves and a loss of muscle tone throughout (this last despite the fact that the falcons remained perched). About 7 percent of the time spent asleep, the EEG showed an activated pattern with clusters of eye movement, which the researchers concluded was REM sleep. This study supports two interesting ideas about the evolution of sleep behavior. First, the fact that falcons have more REM sleep than chickens supports the hypothesis that it is safer for predators to enter this state than it is for prey. Second, the fact that the falcon has muscle atonia without falling off its perch indicates that an animal can adapt sleep to the conditions of its habitat. I will return to both of these generalizations when I consider the variables affecting mammalian sleep.

A PHYSIOLOGICAL FOSSIL: SLEEP IN THE ECHIDNA

Because it happened to live on geographically isolated islands in the Australian Pacific, the nontherian, egg-laying mammal echidna escaped extinction. In the absence of more aggressive placental therians, the echidna had to compete only with the relatively unthreatening marsupial branch of that same family. The American physiologists Truett Allison and Henry van Twyver imported some of these relics from Australia to New Haven in order to test the hypothesis—then popular—that all mammals have REM sleep. Safe in a Yale University laboratory, these burrowing beasts hunkered down deep in their cage bedding and slept for 12 hours a day. But not a single sign of REM sleep was observed. Rather, the echidna's quiescent behavior was associated with an EEG pattern like that of the non-REM sleep of reptiles. This was a surprise that biologists still find intriguing. Francis Crick has recently suggested that the echidna doesn't need REM sleep because its exceptionally large cortex

This short-beaked echidna (Tachyglossus aculeatos) lives in New South Wales, Australia. It constitutes a living fossil record of our nontherian ancestors, whose venerable age is shown in the figure on page 51. The echidna is the only mammal yet studied in the laboratory that has no REM sleep, a lack perhaps reflecting its primitive nature.

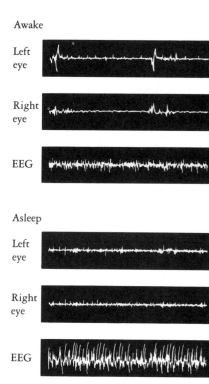

Awake

Left
eye

Right
eye

EEG

Asleep

Left
eye

Right
eye

EEG

The echidna has recordable eye movements when awake (upper traces) but none when asleep (lower traces). The sleep EEG shows high voltage, slow waves typical of non-REM. No brain activation is observed in echidna sleep.

allows it to deal in another way with whatever cognitive and neurophysiological problems are solved by REM sleep.

Allison and van Twyver wanted to establish why the echidna lacked REM sleep. Was the echidna REM-less simply because it was so primitive? Probably not, because the opossum, equally primitive (though marsupial), had plenty of REM sleep, up to 20 percent. Was it because it lived underground? Those subterranean sleep experts, the moles, said, "No, probably not," because they racked up two hours of REM sleep a day—or 25 percent of their total eight. (Both the percentage of REM sleep and the electrophysiological signs found in moles were found in humans.) Allison and van Twyver concluded that the absence of REM sleep in the echidna probably depended in some way on its nontherian mammalian state, but they were unable to establish how.

THE ECOLOGY OF MAMMALIAN SLEEP

Whether or not the brains of diverse species of mammals are considered more similar to one another than different, they are certainly more similar than the sleep of their owners. In fact, there is almost as much variation in sleep *within* the order Mammalia as there is *across* the orders of vertebrates. Several factors contribute to this variability. Variation in body (and brain) size and metabolic rate are some of the intrinsic differences that affect sleep; prey-predator status and dwelling place are among the ecological variables.

Genetic factors: body size and metabolic rate

The two major descendants of the reptiles—the mammals and the birds—have separately evolved two related functions: they both maintain constant body temperature, despite marked fluctuations in the supply of heat, and they both sleep. The ability to maintain body temperature allows an animal to remain active regardless of what the environmental temperature is. This ability gives mammals and birds an obvious advantage over reptiles, which are sluggish in very hot or cold weather because their metabolic rate depends on ambient temperature. But it means they must find ways to keep warm when temperatures are low.

The need to keep warm is particularly great in small animals, because, having a relatively high ratio of body surface to body mass, they radiate heat more rapidly than do larger animals. Small body size therefore contributes to the risk from exposure to cold. We know this intuitively—and perhaps from bitter experience—in raising our own young. Being smaller, as well as immature and physiologically unstable, human infants are more susceptible to cold than we adults. We dress them warmly, even to the point of swaddling, and especially when they sleep.

To generate more heat, physiological systems in small animals speed up and thereby raise metabolic rate. Because more energy is used, the animal must consume large amounts of food or find ways to conserve energy. I have said that hibernation is an adaptive behavior in small animals, allowing energy to be conserved. Non-REM sleep is an adaptive behavior for the same reason: it is hypometabolic; that is, it has low metabolic cost. So we would expect small animals to sleep longer. However, small animals are also especially vulnerable to predation, and that particular risk is best avoided by sleeping as little as possible. Small size actually creates the need to adapt sleep behavior in two contradictory ways to reduce two different risks. The effects of an animal's status as prey or predator are considered in the next section.

As shown in the graphs on this page, increased metabolic rate means that everything goes faster in small animals: heart beat, respiratory rate and, not surprisingly, the frequency of the sleep cycle. Thus, the period length of the non-REM/REM interval is directly correlated with body size, regardless of how much time the animal actually spends asleep. The rat has a non-REM/

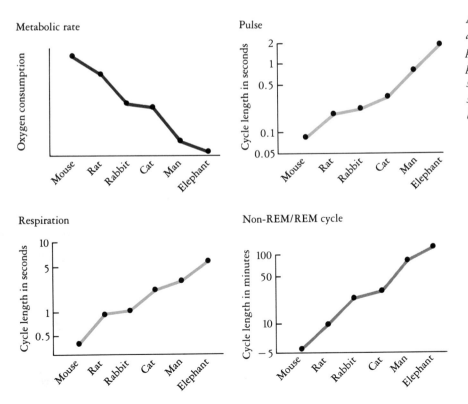

As metabolic rate decreases with an animal's size, there is an increase in the period length of such physiological cycles as pulse, respiration, and non-REM/REM sleep. "The larger, the slower" and "the slower, the longer" seem to be basic rules of biological design.

REM cycle length of 6 minutes, the cat of about 30, the adult human of about 90, and the elephant of 150! As you will see in Chapter 4, this rule applies within species as well as across species: the human infant has a cycle of about 45 to 50 minutes, just half that of its parents. Because the cycle is shorter, small mammals have more REM periods per hour than large ones and these more frequent REM periods tend to be of shorter duration.

Predator and prey

We humans, living in safe, heat-regulated houses, have difficulty appreciating the hazards of sleeping out of doors. We may have to recall childhood summer camp overnights to put ourselves closer to the experience of most animals sleeping in the wild. Their fur is their sleeping bag, and their nest, if any, is their tent. And in our snug homes, we are so preoccupied with the psychological predators that excite our fears and anxieties that we may have difficulty realizing that many animals sleeping out must contend with exposure not only to the elements but also to the ultimate danger of becoming some other animal's dinner or breakfast.

In this herd of wildebeests, some animals graze while others rest. Rotating active behaviors with passive, soporific ones among different herd members may enhance survival. At all times, some animals are available to act as sentinels that can warn the group of predators.

Unprotected animals trying to avoid becoming a predator's meal simply can't afford to sleep long. They must stay alert as much as possible in order both to spot enemies and to run away quickly. Their need to be alert suggests what the ecological value to vulnerable species might be of having a brain-activating system like REM built into their sleep. REM may prevent vulnerable animals from remaining asleep too long at any one time. This "sentinel" theory was elaborated by the American psychiatrist Frederick Snyder in the early 1960s. Before it was known how sleep-shy large herbivorous mammals were, Snyder proposed that by periodically waking up in sleep, each at a different time, the members of a group of zebras dozing in the savanna could keep their collective eye peeled for the hungry lion.

The lion has been called the king of the beasts because of its position at the top of a prey-predator pyramid. Although this large and aggressive cat has not yet volunteered for sleep lab studies, I have been told by my colleague Donald Fawcett that they are not only champion eaters but champion sleepers. After consuming their fill of gazelle or antelope meat in an hour or so, lions then sleep more or less continuously for two or three days, generally out in the open.

The ability to sleep at length and out in the open possesses certain clear advantages for the lion. Sleep, because of its attendant decrease in heat produc-

Frederick Snyder was so fond of animals that he raised a chimpanzee, like the ones shown playing with him here, in his family home in Bethesda, Maryland. This naturalistic bent led him to conceptualize REM sleep as a third major organismic state and to suggest that it might function as a physiological alarm clock allowing sleep to be punctuated, if needed, to monitor the environment.

Dead to the world are this sleeping lion and lioness. So safe are they from predation that they sleep belly up, in broad daylight, and fully exposed in their savanna habitat.

tion, is not only a way for the lion to conserve calories (and thus the ability to generate heat) when the weather is cold; by reducing heat generation when the weather is hot, sleep is also a good way to avoid overheating. In the lion's case, sleeping in the open facilitates radiation of body heat—obviously of particular importance when it is hot.

We don't need to go to Africa to appreciate the sleep privilege enjoyed by carnivorous predators. Regard the domestic cat and recognize its capacity, when taken to the country, to revert to its hunter instinct. All a cat needs to do is sense the presence of prey—be it mouse, squirrel, or bird—and it will stalk with a stealth both wondrous and terrifying, pounce, and kill. Following the hunt, the cat will typically ape his lion relations and go to sleep, whether he eats the mouse or not.

We could say that our cats sleep a lot because they have nothing else to do. They are "bored." And that is, in part, true. We protect them from any real or imagined predator, and we keep them warm. But does boredom really explain why cats sleep as much as 16 to 18 hours a day? When we put together the cat story and the lion story, we recognize that it is equally true that these creatures sleep because they have easily and efficiently achieved their biological goals: survival and procreation. Why waste energy killing more mice—or more gazelles—than they could eat in a week?

Dwelling place

Squirrels have two ways of avoiding cats that have repercussions for their sleep behavior. One is to keep a sharp lookout. Consider the unsettling nervousness of the squirrel on your garden fence. Even when he has a nut in his tiny forepaws and is chewing away, he is twitching, bristling, jumping, and above all, looking at you looking at him. Make one false move and that squirrel will skitter away. It is almost as if his constant movement was a way of keeping his motor system ready for a fast start, as well as a way of enhancing vigilance. In this sense, a squirrel is like a Jimmy Connors—alert and fidgety, the sooner to anticipate and meet the flashing serve of John McEnroe. These defensive adaptations must add to the squirrel's already high energy costs and thus to the squirrel's food and sleep requirements.

Because of fast reflexes and low body weight, the squirrel is able to put his nest in tree limbs that are too high and too small for most predators to reach. Thus, prey status helps determine the location of the squirrel's sleep site. By making the nest both warm and out of the way, the squirrel is able to sleep efficiently and safely.

Differences in dwelling-place safety can account for very different sleeping patterns in animals that are similar in size and metabolism. Allison, in discussing this point, contrasts the bat, a flying mammal who sleeps well, with the shrew, a terrestrial creature who sleeps hardly at all. Both are small, and in both activity has a very high metabolic cost. Nonetheless, whereas the shrew scurries around in an almost constant search for food and sleeps little, the bat sleeps 20 hours a day, safely suspended from the ceiling of a cave by a cliplike claw.

Not all supersleepers are predators, nor do they all dwell far above the ground. The 13-lined ground squirrel and the hamster are good examples. Neither is predatory, and both are subject to predation. Yet because their burrows and nests are so safely situated, they do not need to monitor their environment and hence, like the bat, can afford the luxury of metabolically conservative sleep. Both these species are also hibernators.

And so is the woodchuck, who adds another stylistic detail to the adaptive repertoire of sleep. Perhaps because he is so large, but also because he is

Safe from cold and predators, two mice sleep peacefully in their nest.

so indolent, this marmot emerges from his burrow to feed only twice daily—at dawn and at dusk—and then, with impressive furtiveness and efficiency, eats enough grass in 20 minutes to keep going all day or all night, or even all winter! The regular dawn and dusk feeding periods are strongly suggestive of a circadian rhythm entrained by the level of light. In their restriction of outings to such short and specific times, these fat and sleek beasts show clearly how circadian and sleep factors may interact to promote both unhindered foraging and a life of enviable leisure. Of course, we don't know that woodchucks sleep all the time in their burrows—they must dig a good deal, for example—but I would be surprised to learn that they played chess, or read *Scientific American,* to while away the hours!

The macaque monkey of Asia is another example of a deep-sleeping nonpredator. Like the squirrel, the macaque sleeps in the dense foliage of treetops and at night, when its predators are rare. And like the squirrel, the macaque is light and agile and can thus sleep on high, small branches, which will not support his hunters' weight. In the laboratory, macaques show a nearly human pattern of sleep, with periods of EEG slow waves interspersed with REM sleep epochs.

The African baboon, though larger and a better fighter than the macaque, is a fitful sleeper who rarely enters the REM phase, even in the laboratory. In the wild, he must avoid his chief predator, the leopard, who is a good climber and active at night. And because the leaf cover is poor in the scrubby trees of the savanna, the baboon is easily seen silhouetted against the

This male chimpanzee sleeps in a hammock of vines and branches high above the jungle floor in Gombe, Africa. Because the branches are light, they would translate an intruder's presence into an awakening stimulus.

sky, even in the highest branches. So fierceness is not all, even when combined with size. Insomnia emerges in animals and in man as an ecologically determined symptom.

LONGEVITY AND SLEEP

Because shrews live only 2 years while bats, though no larger in size, live 18, Truett Allison has suggested that the capacity to reduce metabolic rate in sleep may contribute to longevity. This is potentially an important theory, especially if valid for individuals within a species as well as for cross-species comparisons. Edna St. Vincent Millay evidently believed some such theory when she penned this stanza:

> *My candle burns at both ends*
> *It will not last the night*
> *But, ah my foes and oh my friends*
> *It gives a lovely light.*

Whether insomniacs would agree with the poet on the compensatory advantages of high energy expenditure is another question. Still, we would like to know, "Is it true? If I am energetic like the shrew, will I die young? And if I learn to relax or practice meditation, will I live longer?"

At their University of Chicago sleep laboratory, Harold Zepelin and Allan Rechtschaffen examined data from 40 species in 12 orders of mammals and correlated a variety of sleep measures with metabolic rate, brain size, and life span. They confirmed an association between high sleep quotas and high metabolic rate, small brains, and short life span. But the correlation between life span and sleep quota was actually negative ($-.81$) when the metabolic rate variable was partialled out, suggesting that sleep length itself does not determine longevity at a species level.

THE BEASTS OF THE FIELD

The hoofed animals, or ungulates, have already been mentioned as the unprotected prey of carnivores and hence as generally light sleepers. In addition to the need to guard themselves from predators, ungulates have another reason for being light sleepers. That reason is the diet of these herbivorous creatures. Because grass is a very poor food, requiring a specialized digestive system, hoofed animals need to eat a lot of it, and they must feed a great deal of the

Percentage of day spent awake, in non-REM sleep, and in REM sleep

	Awake	*Non-REM*	*REM*
Pig	46.3	46.4	7.3
Cow	52.3	44.5	3.1
Horse	80.0	16.7	3.3

time. The behavior of hoofed animals reflects their need to keep nourishing themselves as clearly as their multiple stomachs. Thus, the cow has a cycle of grazing while the food is ingested and repose while the food is digested. The process is called rumination or, in the vernacular, chewing the cud. The cow's feeding rhythm is a specialized example of Kleitman's basic rest-activity cycle.

The domesticated ungulates are important to us because they provide protein in the form of meat, milk, and cheese, in addition to being used for transportation and work. We have an unusual relationship to these animals in that we have become in a sense both their protector and their predator. The effects of this dual relationship on the sleep of farm animals are enlightening and, fortunately, have been studied by Yves Ruckebusch, a French veterinarian at the University of Toulouse. Ruckebusch's findings regarding the sleep cycle in three different farm animals are summarized in the table on this page. Because laboratory recordings of these animals in the wild are not available, a definitive comparison between domestic and untamed animals within the same species is not possible. Yet Ruckebusch's observations suggest that domestication may enhance sleep by solving the problems of food finding, temperature control, and protection.

These results in barnyard animals again reveal the flexible nature of mammalian sleep and its sensitivity to different environments. It should perhaps forewarn us about seeking rigid rules about sleep behavior in our own species.

INTERACTION OF ECOLOGY AND CONSTITUTION

In order to quantify the contributions of environmental and intrinsic forces on sleep, Allison and Cicchetti performed a statistical analysis on data from 39 species in 13 orders of mammals. Several ecological variables were calculated on 5-point scales: the degree of predation, exposure to danger during sleep, and overall danger. For example, an animal like a fox, which sleeps in a den (exposure index, 0-1) and is not hunted (predation index 1), would have a low overall danger score and, on the basis of ecological factors, would be expected to sleep well. The constitutional variables measured were body weight, brain weight, life span, and gestation time.

The constitutional variables were found to correlate positively with one another but negatively with non-REM sleep. In other words, a small body and a small brain predict high levels of non-REM sleep, short pregnancies, and short lives. As previously pointed out, the critical factor in this set of relationships is metabolic rate, which may in turn be considered the energy cost of staying warm.

The extent of the danger of sleeping was also negatively correlated with sleep, especially REM. The greater the predatory danger, the less REM sleep.

In discussing their data, Allison and Cicchetti suggest that for small animals, the high amounts of sleep are due to its energy conservative properties, while, for large animals, the low amounts of sleep may be the combined result of the herbivorous foraging pattern and the high vulnerability to predation of animals too large to hide.

ANIMAL CONSCIOUSNESS AND DREAMING

Do animals dream? Since so many mammals *do* have sustained periods of REM sleep, what can we infer about their experiences during REM sleep? The question is, of course, moot if posed in human terms. When we ask each other, "Did you dream last night?" we really mean, "Did you notice and, if so, can you render a report?" We would not anticipate a dream report, I dare say, even from the most linguistically gifted chimpanzee. Nor does it appear, based on what animals can tell us, that whatever consciousness they possess is organized in the narrative form that ours is. For our dreams are a clear example of a narrative form of consciousness, as are many of our waking reflections. We cannot imagine thinking without telling ourselves things; plans for the day, reactions to people, abstract analyses, and, of course, fantasies all involve propositional language. We talk to ourselves all the time.

But many elements of consciousness operate at a nonnarrative, nonverbal level. We perceive the world and recognize salient features, like familiar versus strange, or hostile versus friendly, by using parts of our brain-mind that are nonverbal. This perceptual-emotional level of consciousness is probably shared by other mammals. Certainly mammals see, recognize, learn, and emote. And memories of these experiences must be stored in the brain. So if their brains are activated in sleep, I don't see why this part of their consciousness should not also be activated.

When I tried to explain my research on REM sleep and dreaming to my Vermont neighbor, Marshall Newland, who is an inveterate observer of animal behavior, he immediately said, "Oh, you mean when the dog dreams that he is chasing rabbits, he whines and makes running movements with his feet?" Yes, that is exactly what I mean. Motor action begets scenario construction, which begets the selection of motor action to fit the scenario. After all, I need not say to myself I am running to *know* that I am running. If a dog's brain commands running movements, I don't see why he shouldn't somehow perceive movement and then associate that perception with a visual and perhaps even an olfactory image of that familiar object of his chase, the rabbit.

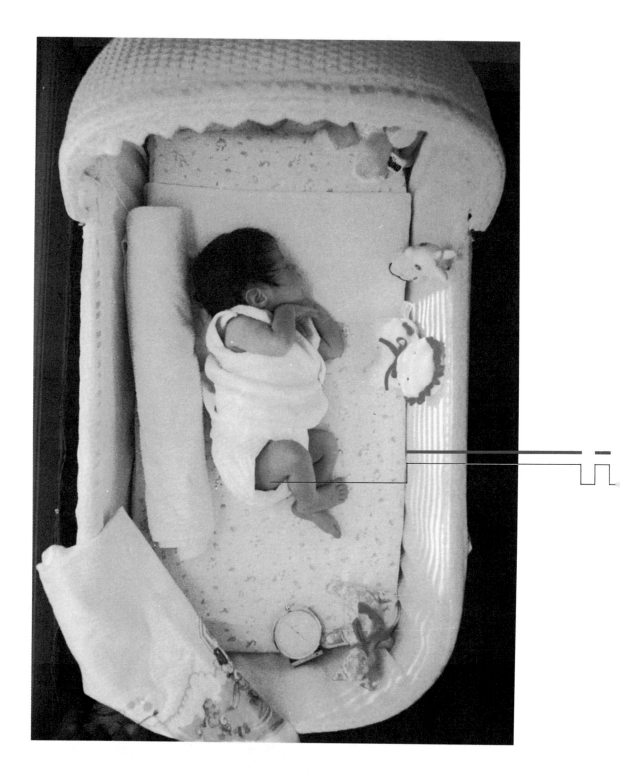

4

The Development of Sleep

It was long held that ontogeny, or individual development, recapitulates phylogeny, or evolutionary development. If that were so, then we would predict, in view of the evolutionary picture presented in the previous chapter, that non-REM sleep arises earlier in human development than REM sleep. Yet the facts run counter to the prediction: REM sleep is the earliest recognizable sleep state, while non-REM sleep develops much later.

These surprising findings are best explained by the fact that brain structures develop at different rates. The brainstem structures supporting REM sleep mature early in intrauterine life, whereas the systems necessary to non-REM sleep mature only after birth. The corresponding changes in our brain and sleep capacities again underline that sleep is of the brain and by the brain. Moreover, they suggest a way in which sleep may be for the brain: sleep may somehow contribute to the brain's construction.

Immature mammals, such as this infant, sleep much more than their parents. The chart above illustrates the cycles recorded during three hours in the sleep of a three-month-old child. Much more of infant sleep than adult sleep is devoted to the REM phase (indicated by the red bars). These findings suggest that REM sleep may have a role in brain development.

SLEEP AT BIRTH

The chart on this page shows how the proportion of the day devoted to REM sleep, non-REM sleep, and waking changes throughout life. The human infant born at full term sleeps for 16 to 17 hours a day; half of this sleep is REM sleep. At sexual maturity 12 to 15 years later, the individual will sleep for only about 8 hours, and only one-quarter of this sleep will be REM sleep. Thus, in absolute terms, the human in infancy has more than four times as much REM sleep as in adolescence. What is the explanation of this dramatic difference? One view is that the decline in REM sleep is the result of brain maturation. And that is almost certainly the case, as you will see. But it may also be that REM sleep is so much greater in infancy because it favors brain development. This is an exciting hypothesis, and in this Chapter I shall examine some of the supporting evidence.

The other half of the newborn infant's 16 hours of sleep resembles adult non-REM sleep in that the infant shows little motor activity. But the EEG patterns do not reveal their full electrical complexity, with spindles and delta waves, until later, when, presumably, the reorganization of the cerebral cortex in the first few months of life has made this complexity possible.

The relative proportions of each 24-hour day that are devoted to wake, REM sleep, and non-REM sleep change dramatically over our lifetime. Exactly how and when these states develop in early uterine existence is not known (dotted lines), but data from premature infants suggest that REM sleep is almost all of life at 26 weeks of gestational age. After 26 weeks, waking increases progressively and inexorably until death.

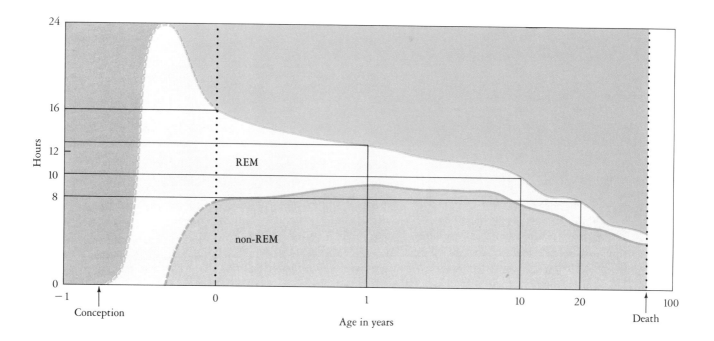

Not only do newborns have more REM sleep than adults, but the behavioral manifestations of REM sleep are exaggerated. Probably because the brain's inhibitory systems are relatively immature, infants show much more dramatic muscle activity in REM sleep than will later be observable. Thus we witness—and delight in—a surprising range of facial expressions, billings and cooings, stretchings or clutchings of the hands, and reactions of the whole body resembling startle responses. Watching a sleeping newborn infant is the easiest way for you to *see* REM sleep and appreciate the brain activation involved. And it is the easiest way to imagine the sequences of motor acts upon which our dream scenarios are laid.

The highly patterned motor behavior that is seen in the REM sleep of newborns has important implications for our understanding of the functions of sleep (as will be discussed in Chapter 9). It seems clear that very early in life the brain achieves a degree of organization that makes it capable of generating elaborate motor behavior. Motor behavior therefore appears to become possible long before it is consciously commanded (as it will later come to be in the toddler's walking). If REM sleep is able to generate movements well before they are needed, then perhaps they serve another purpose. I will propose later that the brain activation of REM sleep contributes to the development and maintenance of our sensorimotor competence.

Another major difference between infant and adult sleep is the tendency for babies to have REM episodes at sleep onset, which has already been noted as having helped Aserinsky to recognize that state in children. The tendency also makes it easy to observe the REM state in babies, because that state so often follows predictably upon a feeding. When a baby finishes suckling, a loss of muscle tone definitively signals the onset of sleep; then the eye movements are directly visible, along with the other motor signs. As nap studies have shown, the human adult may also enter REM sleep directly at sleep onset during the daytime.

THE BRAIN AND BEHAVIOR IN UTERO

Sleep patterns at birth and thereafter follow from changes that have occurred in the womb. To convey the development of sleep and the non-REM/REM cycle, I therefore need to begin by describing changes in the fetus. I will then come back to describe the progression in sleep patterns from infant to adult.

As discussed in Chapter 2, our biological rhythms are expressed through a sequence of states, each of which is composed of a relatively invariant set

The automatic and highly organized movements that are seen in REM sleep begin long before birth, as mothers kicked by a lively fetus can attest. Using high-frequency sound waves as an energy source and a radarlike computer system to read their echoes, Jason Birnholz has documented these REMs and complex facial expressions in utero.

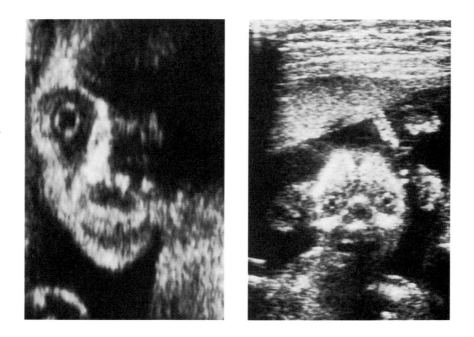

of physiological conditions. The characteristics that define a state include eye opening and closure, respiratory pattern, and the presence or absence of eye and body movement. Thus, the emergence of states necessarily depends on the development of the nervous system. Once motor neurons have appeared at 6 weeks after conception and sensory neurons at 8 weeks, the embryo is able to turn its head in response to a stimulus. From 8 to 20 weeks, motor control becomes more precise, and the area of the body surface responsive to sensory stimuli enlarges. Within this same period, respiratory movements appear and become rhythmic. Thus, by 20 weeks of gestational age, the nervous system is able to carry out both reflex and spontaneous movement.

Much of our knowledge about the development of sleep in the later months of gestation derives from studies of premature infants undertaken by Colette Dreyfus-Brisac and Nicole Monod in Paris. Their work confirms that the motor system is activated fairly early and that the organization of behavior into states comes only later. Infants born at 24 to 26 weeks, who usually do not survive, writhe continuously. This early state of activation will later become REM sleep. Electrical activity of the brain at this point is flat and sporadic.

Between weeks 29 and 32, brain electrical activity becomes continuous. The first behavioral signs that distinct physiological states are present occur at

28 to 30 weeks, when movement is periodically interrupted by quiescence. At 32 weeks, periods of quiescence reoccur at regular intervals, and REM sleep occupies 50 percent of each fetal day. EEG recordings made during the quiet periods show "bursts" similar to those seen in non-REM sleep; these bursts are separated by long flat periods. From 29 to 40 weeks, the duration of the bursts

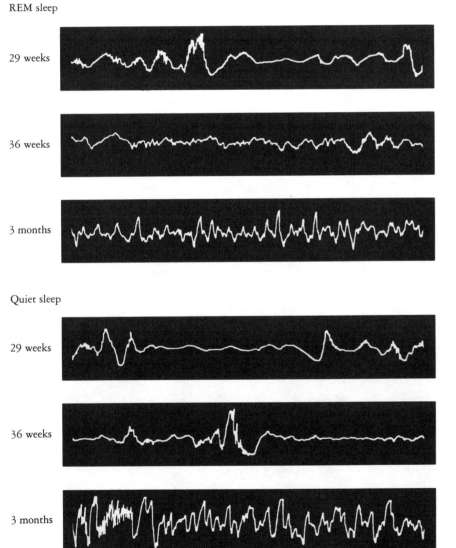

REM sleep

29 weeks

36 weeks

3 months

Quiet sleep

29 weeks

36 weeks

3 months

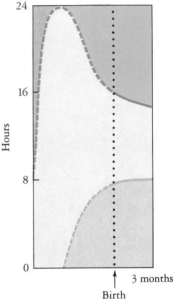

As the brain matures, its electrical activity becomes increasingly complex. This evolution is particularly dramatic in non-REM (or quiet) sleep, where an initially flat and featureless EEG tracing (29 weeks) comes to be interspersed (36 weeks) and then dominated (3 months) by high-voltage slow waves and spindles. But even in REM, which tends to be low voltage and fast at all ages, the EEG shows a similar progression to higher-voltage, slow activity.

doubles, while the length of the flat periods halves. The EEG recordings on page 75 illustrate both the increased activation of REM sleep from 29 weeks to 3 months and the growth of slow waves in quiet sleep during the same period.

SLEEP DEVELOPMENT DURING THE FIRST YEAR

Although infants have some innate capacity to concentrate sleep in one part of the day, this capacity must be developed in the weeks after birth. The entrainment of this rhythm becomes a major task for the parents and infant during the first three months of life. Successfully completed, it allows both to spend most of the night asleep and to concentrate most of the waking periods in the day.

The process is aided by the infant's increasing tendency to consolidate several non-REM/REM cycles into successively longer sustained sleep periods. The newborn infant's patterns are best characterized as serial episodes of sleep punctuated by waking. Gradually, there is a shift to a polycyclic pattern, in which sleep bouts of 5 to 6 hours occur at one or more times during the day or night. By 3 or 4 months, the infant may sleep as long as 8 to 9 hours at night and may nap at predictable times during the day. As the infant's sleep becomes more regular, the wake periods become consolidated as well. The longest wake period, only 1 or 2 hours at birth, increases to 2 or 3 hours by the end of the first week and to 3 or 4 hours by the end of the first month.

As the cyclic structure of sleep becomes clearer, the proportion of each cycle devoted to REM sleep drops. It falls from 50 percent at 3 months to 33 percent at 8 months, with the proportion devoted to non-REM sleep increasing correspondingly. The drop in REM sleep from before birth through 8 months is clearly shown by the sequence of sleep cycles on the facing page. Moreover, although the sleep cycle duration does not change significantly, the

REM versus non-REM sleep in the first year

Age	Duration of sleep (hours)	REM			Non-REM		
		Amount (hours)	Percentage	Net change since birth (hours)	Amount (hours)	Percentage	Net change since birth (hours)
Birth	16	8	50%	—	8	50%	—
8 months	13	4.3	33%	−3.7	8.7	67%	+0.7

total time spent asleep falls from a postnatal level of 16 to 17 hours each day to 14 to 15 hours at 4 months and 13 to 14 hours at 8 months.

A simple calculation shows that all 3 hours that have been dropped in the first 8 months of life must have been REM sleep and that, in addition, almost another hour of REM has also been lost by a shift within sleep to non-REM sleep. The table on the facing page summarizes these changes. As non-REM sleep becomes more dominant, it increasingly resembles adult non-REM sleep, with more elaborate EEG wave forms and spindles.

REM sleep periods become shorter and non-REM (quiet) sleep periods become longer as infants grow from prematurity (34 weeks) through birth (40 weeks) to 8 months of age.

34 weeks gestational age

36 weeks gestational age

40 weeks gestational age (birth)

3 months

8 months

10 minutes

BRAIN DEVELOPMENT AFTER BIRTH

REM sleep depends on the activity of reticular neurons in the pons, and these are already relatively mature at birth. However, it is only between 1 and 3 months of age that the cells in the cortex become extensively interconnected and that their connections to and from the thalamus are established. It is these changes that make possible the development of EEG spindles and the slow waves of non-REM sleep. Chapter 6 will have more to say about the brain processes underlying REM and non-REM sleep.

The brainstem, meanwhile, has not completed its own development. The neurons in the noradrenergic and serotonergic systems must still establish contacts both locally, in the brainstem, and remotely, in the cortex. Recall from Chapter 1 that these are the neurons that are active during waking and non-REM sleep but that stop firing in REM sleep. One function of the noradrenergic and serotonergic systems is inhibition, and the declining levels of REM sleep during the first year are in part the result of an increase in aminergic inhibitory control. It is also known that during the same period neurons become less sensitive to acetylcholine, the neurotransmitter that enhances REM sleep. Thus, complementary changes in the two neurotransmitter systems lead the amount of REM sleep to shrink.

THE DEVELOPMENT OF SLEEP IN ANIMALS

In other mammals, such as the rabbit, the rat, and the cat, sleep development follows a course similar to that described for the human infant. In all three species, REM sleep decreases rapidly after birth. Depending on their level of maturity, the infant animals may soon also show an increase in slow wave sleep as well as an increase in the time spent awake. However, mature sleep develops much faster in these smaller animals.

At birth, the rat and the rabbit are very immature, corresponding roughly to 26-week-old human premature babies. Significantly, in the early postnatal period their bodies twitch constantly for about 70 percent of the day. There are neither eye movements nor neck muscle activity, and the EEG is intermittently active but undifferentiated. This primordial state evolves rapidly in the first 7 to 10 days. Eye movements appear—first singly, then in clusters—indicating an agitated sleep state homologous to REM sleep. The EEG becomes more continuous and elaborated. Neck muscle activity of the kind necessary to control posture is seen, and a state of quiet or non-REM sleep becomes possible for the first time. Kittens are born slightly more mature since they have eye movements, but otherwise follow the same pattern.

THE DEVELOPMENT OF SLEEP

Unlike the rabbit and cat, the guinea pig completes most of its development in utero. All three of the major states are already clearly differentiated in these newborn guinea pigs, and the states are present in adult form and amount. The work of Daniele Jouvet-Mounier reveals that the guinea pig completes its program in the first week by only slight increases in waking (from 55 to 65 percent), at the expense mainly of its already mature slow wave sleep. REM sleep, at an exceedingly low 5 percent level in this herbivore, changes little.

Once the agitated sleep state has assumed a more or less adult form, its daily amount begins to plummet.

There thus appears to be a relationship between the complexity of sleep processes and the motor competence of waking behavior. This relationship probably results from a common underlying developmental process. The reciprocal relationship between the amount of REM sleep and motor competence suggests a more direct functional link between sleep (when motor circuits may be laid down and tested) and waking (when the motor circuits are used).

A BRAIN ACTIVATION THEORY OF DEVELOPMENT

In 1966, Howard Roffwarg, Joseph Muzio, and William Dement, then working at Columbia University, proposed an important role for REM sleep in the development of a functioning brain. They suggested that REM sleep was abundant in early life not only as a result of the immaturity of the brain, but also because it provided the brain with stimulation necessary to development, especially to the development of the visual system. This idea corresponds well with the main strategy that the brain uses in building itself up: it lays down a motor system, activates it, then gradually elaborates and adjusts the action of

Sleep and Development in Kittens

The bars on the charts below record the periods when Maggie's kittens were in (green) and out (red) of their box. When the kittens were first born, they spent all their time in the box, either feeding or in REM sleep. After three weeks they began to emerge as a group at regular times each day for periods of exploration and play. At six weeks the kittens were mature and independent enough to abandon the box.

To study the natural evolution of sleep and the relationship of the changes in sleep to the acquisition of motor skills, I recorded the behavior of Maggie, my family's pet cat, and each of her three litters of kittens. Time-lapse video recorded behavior, while I intermittently assessed the kittens' sleep by direct observation. Comparisons of the three litters showed that there is an impressively reliable sequence of development, with three stages clearly demarcated by changes in sleep and motor skills.

Phase I (birth to day 17). During this phase, the kittens never left their box and Maggie's absences were brief. The kittens slept, except when Maggie roused them to feed. About 80 percent of the time asleep was spent in REM sleep, and the rest in "quiet" or non-REM sleep. The kittens were at first capable of little movement besides suckling. This phase was essentially one of static dependence.

Phase II (day 17 to day 28). Predictably on day 17, the now mobile and open-eyed kittens had their first outing. This and subsequent outings occurred under Maggie's supervision, and the entire group of kittens participated. The actions of Maggie and

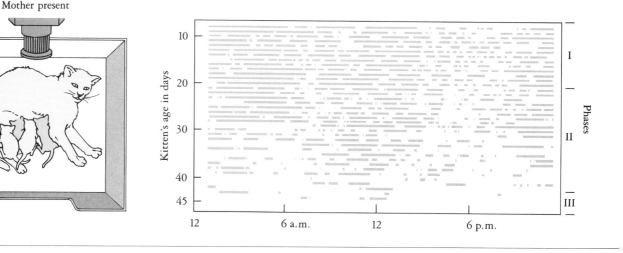

Mother present

the system first by adding the neurons that bring in sensory information and later by building up the internal circuits that are needed for fine control. According to Roffwarg's theory, one way the brain activates the developing visual system is by sending it signals from another part of the brain during REM sleep. These signals might mimic those actually to be received from sensory neurons after birth. Thus, by the time birth occurs, a great deal of functional competence has been achieved by simulating waking behavior during sleep.

the kittens seemed devoted to increasing the kitten's sensorimotor skills, which in fact improved rapidly. The kittens in effect moved from static to dynamic dependence. During this phase REM sleep plummeted, while quiet sleep and waking increased.

Phase III (day 28 to day 40). This phase began with the kittens' first solo sorties and ended the day the kittens abandoned the box they had slept in. The kittens were now able to perform many elaborate behaviors, but the cat family continued to interact as before. The kittens' patterns of behavior and sleep were essentially those of adults, and their waking hours increased yet further.

These observations show how closely linked in time are the changes in sleep and motor ability. They further reveal a sharply demarcated set of phases that determine the behavior of the cat mother and infants. The biological basis of the long-term timing mechanisms that control the phases is unknown, but these mechanisms presumably involve the action of brain clocks and the hormones under their control.

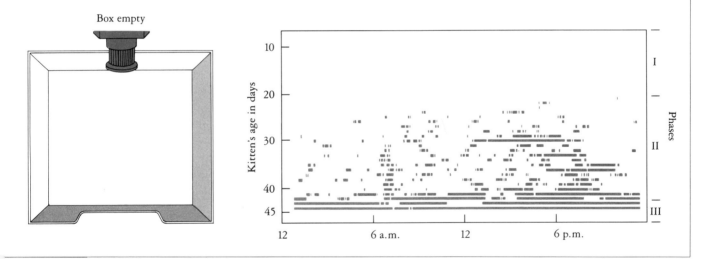

Roffwarg's activation theory focused on vision because experiments with cats had suggested a link between REM sleep and the cats' achievement of visual competence after birth. Once again, a key role is played by the brainstem, which is the seat of eye movement control and the source of the steady electrical activation of the cortex seen in the EEG. In the early 1960s, it was discovered that in REM sleep the brainstem is also the source of pulsatile electrical signals to the visual brain. Some of these signals reach the lateral geniculate body of the thalamus, which in waking is a relay for impulses from

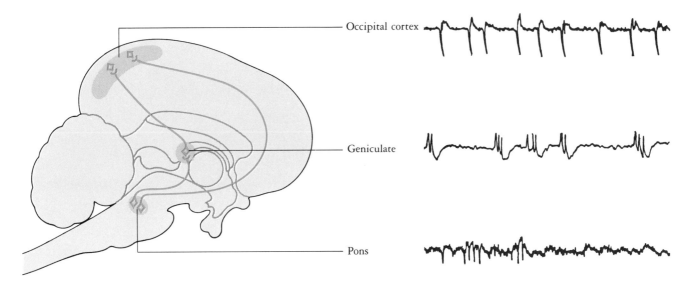

Occipital cortex

Geniculate

Pons

The visual brain stimulates itself in REM sleep via a mechanism reflected in EEG recordings as PGO waves. Originating in the pons (P) from the neurons that move the eyes, these signals are conducted both to the lateral geniculate (G) body in the thalamus and to the occipital cortex (O). Roffwarg has proposed that this autostimulation system facilitates brain development.

the eye to the brain. Other signals go directly to the occipital cortex, the area in the back of the brain where the visual impulses are further processed.

These internally generated "visual" stimuli were called PGO waves, because they were recorded from EEG electrodes in the pontine reticular formation (P), the lateral geniculate body (G), and the occipital cortex (O). Significantly for the question of how they might function in development, not only were these waves an example of one part of the visual brain stimulating another, but the individual pulses carried information as well as energy. Information now known to be encoded in each PGO wave includes both the timing and the direction of eye movement! Although scientists still do not know exactly how REM sleep may favor the development of visual competence, the dramatic changes in both the amount of REM sleep and the functioning of the visual system invite detailed study of the first four weeks of life in the cat.

The important point is that active functioning of the brain is not just the goal of development but is part and parcel of the developmental process. That being so, sleep is an ideal state for some aspects of development to occur, because the brain can concentrate on activating itself as determined by its genetic programming instead of on responding to data from the environment. REM sleep could continue to play a role in the active maintenance of the brain throughout life by guaranteeing the daily use of circuits whether or not they are involved in waking behavior.

LONG AND SHORT SLEEPERS

Sleep patterns continue to evolve across the life span and in the same general direction. That is, sleep capacity declines. By the end of the first year of life, we have already lost half the REM-sleep-generating power we had at birth. In partial compensation, we have become better non-REM sleepers, and our oblivious brains can create quite a remarkable variety of beautiful EEG wave forms. But for the rest of our lives, we will lose more and more of our ability to generate both REM and non-REM sleep until at the end we wakefully contemplate the sleep of death with a strange mixture of anxiety, fear, and relief. This downhill trajectory is not always to be dreaded, however, since people begin their decline from various altitudes, making the fall more or less painful. That is, there are individual differences in sleep length that continue throughout life. The range of sleep lengths seen in individuals is shown by the graph in the margin.

Even closely related species of mice have been found to differ greatly with respect to overall sleep length and depth, and the relative length of the various sleep stages. And within the human species, parents have long observed marked differences in the arousal level of newborns: whereas some are "good" sleepers, others are hyperalert. These traits tend to be lifelong, suggesting that within our species individuals may have a genetically determined propensity to sleep a little or a lot. If so, might such sleep differences somehow correlate with personality types?

The Boston psychiatrist Ernest Hartmann has claimed that short sleepers are energetic, ambitious, and successful, whereas long sleepers are lethargic and passive and tend to underachieve. In the study that produced these conclusions, Hartmann recorded the sleep and evaluated the personality of 16 short sleepers (less than 6 hours) and 23 long sleepers (greater than 9 hours). Subjects, obtained through newspaper ads, were of both sexes; three-quarters of them were age 35 or younger. The mean sleep for the two groups was 5.6 hours for the short sleepers and 9.7 hours for the long sleepers.

Short sleepers appeared to spend relatively more of their sleeping time in Stage IV or in Stage I REM and relatively less time in Stages II and III, as if the extreme stages of brain deactivation and activation were preserved and the intermediate stages were virtually eliminated. Short sleepers could thus be said to sleep more efficiently. In terms of personality, Hartmann characterized these subjects as being "smooth, efficient with a tendency toward handling stress by keeping busy and by denial." They are "What, me worry?" people.

Long sleepers had more Stage I REM and more Stage II non-REM in absolute minutes, as is typical of extended sleep. Hartmann's long sleepers,

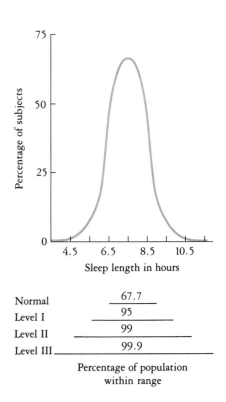

Normal	67.7
Level I	95
Level II	99
Level III	99.9

Percentage of population within range

Like all biological functions, sleep length varies widely. If the number of people who sleep for various durations is plotted on a graph, the result is a bell-shaped curve. Webb's estimation of the chance of falling within 1, 2, or 3 standard deviations from the mean are shown below. Only one in a thousand people falls outside these defined limits, but such extremely short or long sleepers do exist.

especially the males, were "worriers and were clinically somewhat depressed and anxious." But how good is the evidence for Hartmann's assertion?

Wilse B. Webb, a psychologist at the University of Florida, has taken a critical position against Hartmann's findings. He found no significant personality differences in any sleepers except in extreme short sleepers: those who sleep less than $4\frac{1}{2}$ hours do seem to be more energetic.

But now Webb asks: Is their sleep short (on a genetic basis) or is it shortened (by environmental factors)? It is not hard to find people who would like to be short sleepers and who impose sleep restriction regimes upon themselves. I have seen five individuals—all ambitious men and all entrepreneurs or inventors (or both)—who consciously admired Thomas Edison and who consciously emulated his legendary sleep habits. Whether Edison himself was really a short or even a shortened sleeper is unknown, but his claims to have gotten what little sleep he needed by napping under the stairs of his Menlo Park laboratory have led others to believe that they could do the same.

Webb's objection that short sleep may be an act of will, not genetics, is contradicted by the fact that even these highly motivated individuals actually fail to shorten their sleep by their willpower. In fact, they consult sleep disorder specialists because they fall asleep at unexpected, unwanted, and even unacceptable times, such as dinner parties or board meetings. These sleep "attacks" are simply recovery sleep spells, triggered by the self-imposed deprivation experiments of these would-be-Edisons.

As difficult as it is to shorten sleep, it is quite easy to lengthen it, especially for young people. Thus, college students who were paid $5 an hour to produce bona fide EEG sleep doubled their daily output.

Perhaps the only clear conclusion is that there is a rare subgroup of people who constitutionally need little sleep, many of whom appear to take advantage of the fact by getting more accomplished. Not all are enviable, however. The world record short sleeper is an Englishwoman who sleeps 40 minutes a day and who, although she feels fine, is rather bored by her long waking vigil. True short sleepers, for better or worse, appear to be born and not made.

SLEEP IN EARLY CHILDHOOD (AGE 1 TO 5)

Starting from virtually total commitment to sleep at birth, kittens develop an almost complete repertoire of their species' wake-state behavior by only six weeks of age. This rapid progression is a time-compressed version of what we humans do in the first five years of life. The landmark event comparable to Maggie & Co.'s leaving the box at day 40 is a mother and child's march to the school yard when the child is age five. That the kitten becomes independent at a less mature stage and a person at a more mature stage tells us something about the complexity of what it is to be a person as against what it is to be a cat.

Although slower in humans, the development of sleep behavior is just as sure. By age 5, the polycyclic pattern of early life has been completely left behind for a diurnal one. The child rarely sleeps more than 10 to 12 hours at night, and any daytime sleep need is satisfied by a nap, usually in the early afternoon. (A nap at this time prevents the otherwise disabling trough of irritability and dysfunction into which the child would fall from about 4 to 6 p.m.)

In childhood, the brain continues to develop but already has a full range of EEG dynamics, as evidenced by the rapid descent through Stages I through III into deep Stage IV sleep that most young children experience in the first cycles of the night. So obtunded are they that it may be difficult for them to respond either to internal arousal signals, like a full bladder, or to their parents' urgings to awake. The arousal process is difficult, as many parents know, because dissociations between motor, sensory, and cognitive brain functions may occur. That is, the child may respond but not really be awake.

Since social and other demands may increase exponentially during this stage in life, the child must make major psychological adjustments, whose costs may be seen in dissociative sleep phenomena. Thus, completely normal

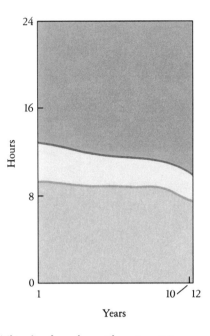

This slice from the graph on page 72 shows the decrease in the daily hours devoted to REM sleep (purple) and non-REM sleep (blue) from ages 1 through 12. Other slices from the graph appear in the margins of the following pages.

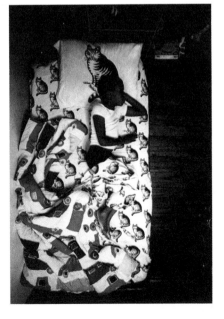

A child of 6 years sleeps from 9 to 12 hours per night. Although the length of sleep is shortening, sleep is very profound, and arousal during the first half of the night may be difficult (see Chapter 8). Some scientists believe that deep sleep serves recovery from the intense physical activity that occurs at this age.

young children may wet their bed, sleepwalk, talk in their sleep, or have nightmares. Parents should convey a sense of normalcy, rather than of alarm, about the many strange things that may occur in sleep. In time these sleep problems will go away.

SLEEP IN MIDDLE CHILDHOOD (AGE 5 TO 12)

During the elementary school years, children experience a relatively steep decline in amount of sleep. Nocturnal sleep drops from about 10.5 to 8.5 hours, and the nap is abandoned. Although children now show a more adult sleep pattern, their brain has by no means completely matured. The social learning and fine-tuning of sensorimotor skills that go on in grade-school children are time-consuming processes.

Children of this age sleep deeply, especially in the first half of the night, indicating that the thalamocortical systems are maturing. These systems, which consist of connections between the thalamus and cortex, generate the spindles of Stage II sleep and the slow, high-voltage waveform of Stage III and IV sleep. Some neuroscientists assume that during sleep and wake alike this part of the brain is busily engaged in storing the abstract information that school offers to the child. The episodes of sleepwalking, sleep talking, and tooth grinding that may be seen in this age group generally occur during deep, Stage IV sleep.

Although they may have brief REM periods at sleep onset, children of this age group typically miss the first emergent (or ascending) REM period and have nothing but non-REM sleep for the first two cycles of the night. The lack of REM sleep early in the night may reflect an increase in the non-REM sleep drive, which is known to be associated with both maturation of the brain and changes in activity pattern such as more exercise.

SLEEP IN ADOLESCENCE (AGE 12 TO 18)

The major landmark of adolescence is puberty and the accompanying changes in body form and size. These changes are the body's response to the brain's elaboration of sexual and pituitary growth hormones, which are released in pulses during slow wave sleep. Sleep does not change dramatically in either quality or quantity, but with growing self-awareness, the young person may become more conscious of his or her sleep needs—and, often, more negligent of them. Adolescence is the time of life when both sex and social interaction become very important. Sexual feelings and sexual scenarios begin to occur in dreams.

Teenage males may begin to experience so-called wet dreams. The fact that ejaculation can occur in sleeping males with no external stimulation and no movement of the body is testimony of the power of REM to mimic the sensorimotor aspects of waking life. For REM sleep is associated with erections in males and erotic dream fantasies in both sexes, and it is almost certainly these elements that collaborate to cause climax during sleep.

This phenomenon may help us to understand the relationship of mind and brain. If sexual fantasies, which can be consciously controlled in the wake state, can produce erections, they must obviously do so through some intermediate brain process. And if I voluntarily induce a sexual fantasy, is my mind not then exerting *causal* control over my brain? I think that it is. But, by the same token, if in REM sleep I suddenly find myself in a state of sexual arousal and hallucinate satisfaction, is my brain not then causing my mental experience? Again, I think that it is. The mind-brain link is a two-way street.

In any case, in adolescence, as at earlier stages, the strain of intellectual, social, and sexual challenges may affect sleep patterns. Adolescents tend to stay up later at night and sleep later in the morning, often staying abed for prodigious sleep bouts on weekends. An obvious implication of the different timing of sleep in adolescence is that the hypothalamic-pituitary mechanisms

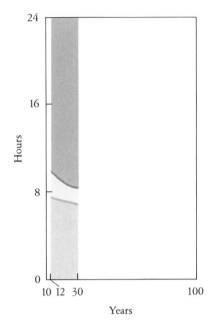

The downward trend in sleep duration continues in adolescence and early adulthood. During puberty, as in this 13-year-old sleeper, non-REM sleep becomes the time when the pituitary glands release growth and sexual developmental hormones, suggesting a new developmental function for sleep.

that are programming sexual development are intimately tied to sleep, not only as effects but also as causes. Scientists might well seek the link between the circadian clock in the hypothalamus and the ultradian clock in the pons through studies of hormone levels in sexually maturing animals.

My own animal studies, in any event, indicate that the sexual hormones may have powerfully arousing effects. In the laboratory, where no male cats were present, I have seen female cat sleep vanish for three successive days during sexual arousal. Thus, our own ability to postpone sleep in the interest of real or imagined sexual opportunities may have a hormonal as well as a psychological basis.

SLEEP IN EARLY ADULTHOOD (AGE 18 TO 30)

Young adults enter the life-long downhill course of declining sleep length and depth without noticing it. There are, however, individual patterns and differences. Early University of Florida studies under Wilse Webb's direction called attention to the fact that the durations of sleep stages were consistent from night to night for any given subject but differed from subject to subject. These studies in effect warn against making generalizations about an age group and argue strongly that the sleep mechanisms governing the kind and amount of sleep each person gets are determined by that person's constitution.

The stages of sleep did not appear in a rigidly fixed order from night to night within or across subjects. Transition from one stage to another was usually smoothest when sleep was deepening but was more abrupt, with frequent jumps, in the upswing toward REM sleep or arousal. As the sleeper progressed from Stage IV to Stage I, he or she might jump directly from Stage III to Stage I or from Stage IV to Stage II. The several stages of non-REM sleep were rarely longer than 10 minutes each in duration, whereas REM sleep lasted up to 39 minutes.

Young adulthood is the time in life when individuals may first notice—and complain about—the way in which their own sleep differs from the norm. By the time a person is thirty, he or she is usually aware of a personal style of sleep and feels more or less comfortable with it. Long sleepers feel comfortable if they themselves and others around them accept their sleep needs. Short sleepers are usually self-satisfied, regardless of the opinions of others.

But courtship, marriage, and childbearing bring a host of variables into the picture; not surprisingly, many of the sleep problems that emerge at this age stem from crying babies or snoring spouses. The timing, depth, and duration of sleep all greatly shape the life of a couple, and people need to learn to sleep together in the literal as well as the figurative sense. Most of our most intimate life with mates is spent asleep, and yet scientists have only the most fragmentary idea of what co-sleep is really like for most people.

SLEEP IN EARLY MIDDLE AGE (AGE 30 TO 45)

It is during the middle part of the life span that most people first notice a shallowing and shortening of sleep. Together with a sense of fatiguing more easily, the decreased sleep length and depth is a sure sign of middle age. The physiological cause of this feeling may be the marked and normal drop in the amount of deep, Stage IV sleep. For those who have suffered from Stage IV−related sleep disorders, this drop comes as a relief; sleepwalking, sleep talking, and nightmares of the pure terror type have abated. But for the rest of us, the loss is unwelcome because we cannot stay asleep as easily: we wake up more frequently, and sleep feels less satisfying. In the morning, we wake up less rested and may find it harder to get our brains in gear.

Since many people change their life-style during this period, it is important to note the range of factors that can make sleep worse. By becoming more sedentary, we may lose the tonic effect of exercise on sleep. By drinking more tea and coffee (to pep us up) and more alcohol (to help us relax), we work in separate ways against the sleep system. If we then add a prescription drug to

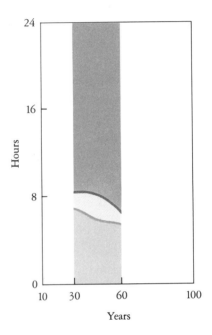

In early adulthood sleep depth always decreases noticeably, and in late adulthood there is sometimes an unwelcome increase in time spent awake.
By age 40, most of us, like this man, have lost all of our Stage IV sleep, the deepest part of non-REM sleep. To the insult of decreasing physical capacity is added the injury of diminished refreshing sleep.

this mix (not to mention the problem of recreational drug use), we will often make matters much worse. Putting on weight may affect posture and respiration in ways that interfere with sleep. And our children come of age, raising the anxiety, aggression, and noise levels of our households to arousing proportions.

SLEEP IN LATE MIDDLE AGE (AGE 45 TO 60)

The climacteric, while most impressively felt by women, also affects men. Loss of hormones has a way of accelerating the reversal of the sleep trends that were seen in adolescence. We want to go to bed earlier; we are much more susceptible to sleep deprivation; and yet we generally sleep even less well than before. Some people learn very quickly to accept the unwelcome arousals that occur after only three hours of sleep. They get up and read, or write letters, or listen to music—and then go back to bed.

Sleep length declines to about 7 hours on average, and Stage IV virtually disappears. In one study, subjects who were 50 to 60 years old had Stage IV sleep only 2.7 percent of the time—quite different from the approximately 20 percent characteristic of 16- to 18-year-olds. With the decline in Stage IV sleep, there is a corresponding increase in the lighter stages of non-REM sleep, especially Stage I (without REMs). We lie in bed neither quite asleep nor yet awake. Our sleep is now decidedly more shallow, in a subjective sense as well as by objective measures. As the prostate gland enlarges with age it becomes more difficult for men to empty their bladders and the need to urinate becomes more frequent—and more urgent. This interrupts sleep as well as perturbing waking life.

SLEEP IN OLD AGE (AGE 60+)

Some of the worst things that happen to most people's sleep during old age are the consequence of changes in social role. We lose our responsibilities. We retire. We give up our houses. We tend to become less involved in vigorous physical effort. All the while, our sleep is still slipping away from us at the same rate. And our days are still 24 hours long. The upshot is that we have more time in which to do less.

In old age, some people come to dread the night because of the loneliness and boredom of lying in bed awake. One good-humored older couple of my acquaintance painted the inside of their bungalow three times in five years. And all this painting was done between midnight and 6 a.m. Not everyone can laugh while telling such touching stories about themselves.

Society needs to concern itself with the problem of how to make better use of its most experienced members. If we cannot engineer a return to the

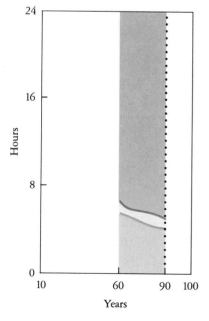

The unkindest cut of all may occur in this last slice of sleep life. Increasing fragmentation and early morning awakenings may perturb the shorter, shallower sleep of old age. Fortunately some of us, like this elderly man, continue to enjoy sound and refreshing sleep.

extended family, where grandparents had a social role and self-respect, we must better construct our communities so that our late life occurs in settings that are more pleasant and more conducive to active participation. In old age, we should derive more satisfaction from the extra time awake that the aging brain guarantees us.

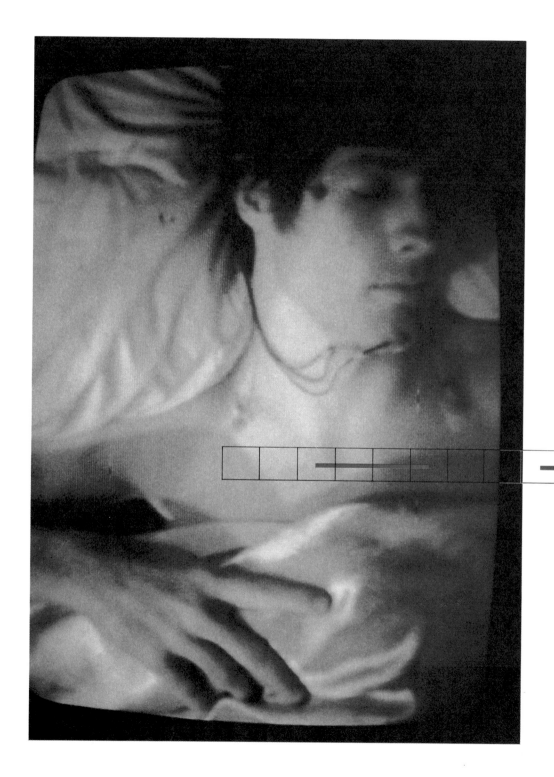

5

Behavior in Sleep

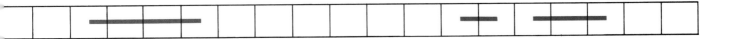

What kind of behavior is sleep? This question, seemingly so simple, hides profound complexity. For example, if an animal is just lying there and isn't *doing* much of anything, can it really be said to be behaving? Isn't sleep the absence of behavior as we usually conceive it? And yet, when we look more closely, we can see that sleep really is behavior, and in two senses: first, sleep involves a quite elaborate internal brain activation, and second, this activation is associated with behavior in the usual sense of the term—outwardly visible action by the sleeper.

Because the brain activation of REM sleep would generate actual behavior if the motor commands were not blocked by inhibition at the spinal cord, we are forced to consider the very real possibility that behavior can be virtual as well as actual—that is, it exists as a set of behaviorial programs in the brain that can be uncoupled from physical action. It follows that we should take

Sleep behavior can be recorded by using time-lapse photography, video, or cinema. The series of boxes above contain the movement profile recorded for the subject shown on page 99, who was photographed at intervals throughout the night. Each box represents a frame from the sequence of photographs; the colored lines indicate when no movement occurred.

seriously the notion that such virtual behavior in sleep—somehow—has a significant function. The distinction between the actual and the virtual carries forward the concepts I advanced earlier suggesting that sleep has both a metabolic role (energy conservation) and a maintenance role (information organization). Although I assume that the two levels are complementary and that they are certainly not mutually exclusive, I concentrate in this chapter on sleep as actual behavior. In Chapter 7, I return to virtual behavior, especially as we experience it in dreaming.

WEBB'S OBJECTIVE BEHAVIORAL MODEL OF SLEEP

Although many people, including professional scientists, believe that little happens outwardly during sleep, some sleep scientists find the outward action of sleep to be significant. One such theorist is Wilse B. Webb, a psychologist at the University of Florida who has proposed a model of sleep that allows scientists to consider how sleep behavior is adapted to environmental conditions. In a recent paper, Webb has emphasized the importance of behavioral analysis as a way of moving sleep studies from a merely descriptive stage to a stage of hypothesis testing. Since such a move is very much a goal of modern sleep neurophysiologists, Webb's model is both timely and welcome. Through its emphasis on adaptation it also brings sleep into the mainstream of Darwinian biology.

Webb's objective behavioral model postulates three major variables that interact to produce sleep. First, the *sleep demand* variable measures the time of wakefulness prior to sleep. According to Webb, our need to sleep increases exponentially with the time spent awake. The longer we stay awake, the sleepier we become. The second variable, *circadian tendencies,* is the timing of sleep within a 24-hour period (discussed in Chapter 2). We feel more tired late in the first night of deprivation than we do the next morning (even though we have not slept). Once we are asleep (and with sleep demand and circadian factors held constant), sleep stages can be predicted as a function of time asleep. These first two variables are both likely to be mediated by the brain. The third variable measures our exposure to *behavioral facilitators and inhibitors* of sleep. These are individual habits (such as lying down or standing up) that promote or hinder expression of the internal drive to sleep. The box on pages 96 and 97 contains an example of how these three variables were manipulated to produce sleep in unfavorable circumstances.

The three primary variables are modified by four factors: (1) differences among species, which are genetically determined tendencies of the brains of kindred animals to be activated or not (discussed in Chapter 3); (2) developmental stages, which are genetically determined changes in sleep across the life span (discussed in Chapter 4); (3) organismic states, which are variables that influence sleep in health and disease (discussed in Chapter 8) and which include disease states (like depression) and the physical effects of drugs (like stimulants and sleeping pills); (4) individual differences within species, which are differences in the tendency of the brain to be activated or not (discussed later in this chapter).

Webb's model defines sleep behavior as sleep onset and sleep termination. I *go* to sleep and I *get* up. Knowing the time of sleep onset and sleep termination allows me to predict such other variables as whether I am asleep or awake, how long it takes me to fall asleep (sleep latency), and how long I stay asleep (sleep maintenance), all of which may be measured as sleep structure using EEG sleep stages. And, all these variables are marked by outwardly observable behavior and inwardly perceptible mental states. The latter include subjective aspects of sleep such as whether I sleep deeply, dream or not, and how rested I feel after sleep. Nesting the variables in this way has an appeal that is practical as well as theoretical: since going to sleep and getting up are easily measured, new approaches to measuring sleep become possible that are both simpler and more natural than the usual laboratory techniques.

MOVEMENT DURING THE FOUR SLEEP STAGES

Now that they can see the lawful way that sleep behavior organizes many objective and subjective variables of interest, scientists can envisage new ways of studying sleep. The most obvious of these is time-lapse photography. Despite having been long available, time-lapse photography was not until recently applied to the study of sleep. Although its use is still very limited, it has already vindicated Webb's contention that sleep behavior is, in and of itself, remarkably informative.

Just as Leland Stanford was indebted to Eadweard Muybridge for proving that horses leave the ground while running, so we sleep scientists are indebted to the contemporary photographer Theodore Spagna for showing that people move in their sleep in a most meaningful manner. Stanford's bet is analogous to the contention of a sound sleeper that, because he went to sleep and woke up in the same position, he had not moved all night long. As I discuss later, photographic studies of human sleepers have shown that even the soundest sleepers shift position at least eight times a night and that the fewer the number of shifts the better people report having slept.

Spagna's interests were aesthetic, not scientific. He studied sleep photographically because he believed it to be "a hidden landscape," whose explora-

Sleeping on Exhibit

My colleagues and I were able to dramatize that if people take advantage of what we now know about sleep, they are able to sleep in even the most difficult of conditions. At Boston's Museum of Science, where the traveling version of our Dreamstage exhibition was installed in 1980, volunteers slept soundly during the daytime in a public setting, exposed to the view of an unknown but large number of their fellow humans!

To achieve this surprising result, we followed the guidance of Webb. We simply made sure that our sleeper's behavior enhanced the three factors influencing sleep: the circadian factor, the sleep demand factor, and behavioral facilitators and inhibitors.

The biological clocks of humans are set for us to fall asleep at approximately the same time each day. We recruited sleepers who, because they worked at night jobs, had already converted their circadian rhythm to sleep by day. This naturally put their sleep in sync with the exhibit's hours of 10 a.m. to 5 p.m. All of our volunteers

Asleep in an art gallery, the Dreamstage sleeper is visible to exhibit visitors through a one-way window. Behind his head is the junction box that conducts electrical impulses from his brain and body to the polygraph recorder in the next room.

worked in hospitals on the so-called graveyard shift, from 12 midnight to 8 a.m.

According to Webb's sleep demand factor, people fall asleep more easily after a long period of waking. With this rule in mind, we regulated not when our volunteers slept but when they stayed awake: they were encouraged to do whatever they wanted in the sleep chamber but to stay awake outside of it. This naturally increased their sleep demand.

Webb's third factor, behavioral facilitators and inhibitors of sleep, covers a wide range of behavioral and psychological characteristics and environmental conditions. We chose sleepers who were young (favorable developmental age), comfortable with the medical equipment we used (favorable mental state), and accustomed to ignoring external stimuli (one subject lived near a fire house and slept through four-alarm fire calls; another never bothered to lower her shades and slept in broad daylight). In addition, the sleepers shared our scientific and educational goals, and were eager to participate. In terms of behaviors that affect sleep, our subjects were exemplary in that all were physically active and none smoked or drank tea, coffee, or alcohol. To further ensure our sleepers' comfort, we permitted those who normally slept with cats or teddy bears, or in idiosyncratic pajamas, to do so. One woman slept with her pet parakeet.

The low hum of a ventilator fan and the low light of our video system served to screen such stimuli as might arise outside the chamber. The screens were important since the exhibit included laser beam projections of the sleepers' brain waves, muscle potentials, and eye position data, as well as "music" produced by passing these same electrical signals through a synthesizer system programmed to reflect the sleepers' internal states.

We controlled the public's behavior by providing visitors with strong incentives to be quiet. These included specific instructions beforehand, a soft rug on which they, too, could sleep, and a request that shoes be removed, which impressed upon people the need for silence. But the most powerful controller of all was the individual viewer's encounter with the sleeping subject. Visitors were simultaneously so curious about sleep and so uncomfortable in the presence of another person's sleep that they naturally inhibited their own behavior so as better to observe, and not be observed by, the sleeper. It is embarrassing to be caught watching someone sleep.

So amazed were some visitors at our subjects' ability to sleep in the exhibit that they assumed drugs must have been given. But the natural sleep they witnessed was instead a testimony to the power of sleep itself.

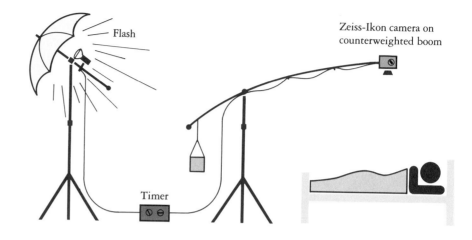

Flash

Zeiss-Ikon camera on counterweighted boom

Timer

In order to make time-lapse color photographs of sleep, Ted Spagna used the set-up shown here. A timer, preset to any desired interval, triggered both the flash and the shutter of the camera, which was mounted on a boom over the sleeper's bed.

Opposite page: The first laboratory study to combine time-lapse photography with standard sleep EEG recording was made on March 25, 1977 at the Massachusetts Mental Health Center. In this contact sheet of photographs taken every 15 minutes, the subject is first seen reading (frames 15–16). After turning the lights out, he then slept soundly, as indicated by the two long periods of immobility (frames 17–20, 21–24, and 28–32). Toward morning (sun-up occurs between frames 32 and 33) movement was more frequent and sleep was lighter. Right: The three long periods of immobility in the study on the opposite page are shown as the aqua bars. The periods correspond to the first three sequences of non-REM sleep as recorded electrographically. Body position often changes before and after the REM periods, shown as the purple bars.

tion would portray significant personality variables. The illustration on this page shows the setup he used to obtain series of photographs revealing sleep behavior throughout the night. Using a 15-minute timer, along with a strobe-driven flash so that he could shoot in color, Spagna simply mounted his camera on a boom over his subject's bed, plugged it in, and went home. Every time the sleeper was "shot," his bedroom was ablaze with light and the camera motor cranked up the next frame with a grinding whoosh. Amazingly enough, some of his subjects slept as well under these rather invasive conditions as did subjects in later studies using invisible infrared light and a silent video camera.

When I first looked at Spagna's studies, I noticed that many were characterized by runs of 3 to 5 frames in a row without apparent movement, followed by a frame documenting a major posture shift. A given posture was never maintained for more than two hours and most were maintained for less than one hour, suggesting the possibility of a relationship between movement and the non-REM/REM sleep cycle. EEG recordings verified this relationship, which proved to be so robust that my colleagues and I were soon able to predict non-REM sleep from the photographic data alone. As we later found, postural immobility of more than 35 minutes is a reliable indicator of non-REM sleep.

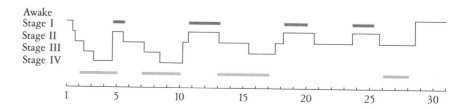

Awake
Stage I
Stage II
Stage III
Stage IV

1   5   10   15   20   25   30

The longest immobile periods within sleep typically began early in non-REM sleep, in the so-called descending phase of the cycle, and ended later in Stage IV or during the transition to REM sleep, the so-called ascending phase of the cycle. The REM periods themselves were also generally immobile, but because they were much shorter, especially in the first half of the night, they could not be recognized from the photographs. The immobility of REM sleep was broken by movement occurring at the end of REM and the beginning of non-REM. These observations suggested a very important generalization: since the body movements often occur during phase transitions, they constitute easily observable behavioral markers of the changing state of the brain. Moreover, because there is a correlation between brain sleep stage and mental state features, body movements may also signal that dreaming will soon begin or has just ended.

As will be explained in Chapter 6, the immobility detected by the photographs has two different underlying mechanisms: in non-REM sleep, movement is unlikely because the excitatory level of the motor system is too low for movement to be commanded, whereas in REM sleep, movement is unlikely because, while the excitatory level is quite high and movements are in fact commanded, the inhibitory level has also been raised so that the execution of the motor commands is blocked. These facts explain how the sleep behavior, sleep stages, and subjective variables result from the activity of the cells in the brain. Although not considered by Webb, the cellular level is essential to complete his model.

Encouraged by the results we had obtained through time-lapse photography, we hoped to find an even simpler, more reliable means of recording sleep behavior. Time-lapse video therefore recommended itself as a promising approach. Instead of taking a continuous movie, time-lapse video stores a sequence of static images taken at whatever frequency is desired. The method has several clear advantages over photography. First, with video we could have more frequent images or even continuous monitoring. Second, far more inconspicuous operation would be possible, as only low levels of illumination were required and the recorder could be placed outside the bedroom. Finally, records that were on video tape could be read immediately, so that we could be sure we were getting data. The results we obtained with time-lapse video confirmed those we had achieved earlier.

By 1982, my colleagues and I knew that sleep was indeed a behavior even in the usual sense of the word as motor activity, and we knew how to use its grossest motor aspect—that is, the posture shifts—as a way of reliably predicting the non-REM phases. But what about REM sleep? Did it have motor signs that could be detected and measured? Because the obvious motor sign of REM sleep is eye movement, we decided to record that movement, which we did directly with a movement transducer in the sleeper's home instead of indirectly with the EOG in the sleep laboratory.

In our first study, we were able to show that eye movements, when properly recorded, could be used to predict the REM periods within sleep without recording the EEG or muscle tone. This might seem like a foregone conclusion, but there are, in fact, an appreciable number of eye movements in so-called non-REM sleep, so their simple absence or presence doesn't discriminate as well as the names of the two states would suggest. The success of such predictions means that ocular behavior alone is sufficient to diagnose REM in sleep.

NIGHTCAP

Once we knew that REM sleep, too, could be diagnosed, Adam Mamelak and I set about designing a new sleep-recording system. The system we developed, called the Nightcap, is portable and yet capable of diagnosing waking, non-REM, and REM sleep. As shown in the illustration on this page, two tennis headbands carry all the hardware and electronics needed to record and transmit the data regarding body movements and eye movements. The former are recorded by an accelerometer at the top of the head, which generates pulses whenever the sleeper rolls over, and the latter by piezoelectric film, which senses lid deformation whenever the eye moves.

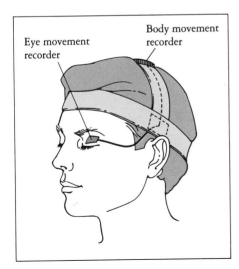

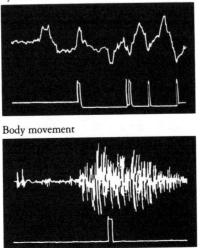

The Nightcap is our simple two-channel device for recording movement at home during sleep. We chose its name because both the body and eye movement sensors are carried in a modified tennis headband, which the sleeper can put on his own head at bed time. Since the development of the prototype device shown here, the sensors, the headgear, and the electronic recording equipment have been streamlined by the Healthdyne Corporation and the device has been tested with hundreds of subjects.

In these time-lapse photographs, the synchronous posture shifts of a couple punctuate periods of highly complementary immobile postures. Their unconscious communication allows the bed partners to share their space efficiently and, we might suppose, comfortingly.

The analytic concept for this two-channel system is simple: If both the body and the eye channels are active, we diagnose waking. If neither channel is active, we diagnose non-REM sleep. If the body channel is inactive and the eye channel is active, we diagnose REM sleep. The fourth possible permutation—body active, eye inactive—does not occur.

To cope with the many variations in pattern observed from subject to subject, an analytic program (or algorithm) tailored to the individual is developed to create the best possible fit between the Nightcap data and standard EEG sleep laboratory measures. We have been able to achieve an 89 percent agreement between Nightcap-diagnosed and EEG-diagnosed wake, non-REM, and REM sleep states on a minute-by-minute basis. Since we can collect many more nights of data under home conditions than we could in the lab, and at a much lower cost, we feel that this error (or difference) margin is tolerable.

SLEEPING TOGETHER

Long before I met Ted Spagna, he had quite naturally expanded his photographic studies to include couples. As Spagna's early studies soon revealed, sleep tends to become synchronized in any two bed partners. That is, the long periods of immobility tend to occur at the same time in both members of a sleep pair.

Most bed partners go to bed at the same time. As a result, they are likely to go to sleep at about the same time and therefore to have similarly timed

non-REM/REM cycles. However, the probability that sleep onset and sleep cycles are synchronized is greatly increased by an interactional factor: each partner is now the most powerful stimulus available to the other. Thus, if the one tends to go to sleep more quickly and the other tosses and turns, the one will be awakened; neither will sleep easily until both do. And if the one who falls asleep also snores, the posture shift by which he or she is awakened may be quite conscious and active. But, whether the awakening is intentional or unintentional, the consequence is the same: the non-REM/REM clock is reset at zero until both partners are "go." Now the cosleep rocket is launched.

As both traverse the non-REM phase of the cycle in parallel sleep orbits, both are likely to enter the transition to REM at the same time. The effect of this paired lightening of sleep is to make the phase transition body movement of the one kick off a nearly simultaneous body movement in the other. Thus, the ascent into Stage I REM is given another strong synchronizing pulse! It may be biologically trivial but it is nonetheless charming to realize that by sleeping together, couples increase the chances of dreaming together.

"TOSSING AND TURNING" OR "SLEEPING LIKE A LOG"?

After looking at Spagna's data, I hypothesized there might be a correlation between the timing and number of posture shifts and the continuity and "goodness" of sleep.

The differences in amount of movement among sleepers were clearly substantial. Thus, while some moved as few as 8 times a night and remained

Many species are even more communal in their sleep habits than are humans. Perhaps to conserve heat and guarantee safety when asleep, these California sea lions are literally piled on top of one another at their beach and slumber party.

immobile for up to two hours, others moved at least 30 times and had fewer and shorter periods of immobility. My colleagues and I calculated an immobility index by counting the number of adjacent frames without movement and dividing by the total number of frames; the index was expressed as a percentage which rose as sleep was increasingly tranquil.

In our studies of over 200 sleepers using time-lapse photography, a perfect score of 100 percent, indicating no movement from start to finish, has never been obtained. Nor have we ever seen a score of 0. Such a score, which would indicate that during no 15-minute period was the sleeper immobile, would presumably be possible only in states of severe agitation.

To test the prediction that the immobility index could predict sleep continuity (by definition) and goodness of sleep (by implication), we studied sex- and age-matched groups of six self-described "good" sleepers and six self-described "poor" sleepers. Each person was photographed for four nights in the laboratory. The results showed that time-lapse photography could indeed discriminate between good and poor sleepers with a high degree of confidence. The immobility values ranged from 47 to 63 for good sleepers and from 22 to 43 for poor sleepers. Hence the scores of the two groups showed no overlap whatsoever.

FALLING ASLEEP

To test Webb's idea that behavior when going to sleep is the outward sign of a brain process, researchers at my sleep lab and I focused our attention on the relationship between posture shifts and EEG changes at the outset of sleep. Video enabled us to pick up more movement than was previously possible and thus to time sleep onset much more accurately. We found that postural immobility reliably *preceded* the first unequivocal sign of sleep (the spindle wave characteristic of Stage II) by a fixed time period, about 7.5 minutes. The tossings and turnings before sleep onset and the relative stillness afterwards are contrasted in the bottom sleep cycle chart on this page.

The finding that immobility precedes sleep reaffirms our common-sense recognition of the role of postural recumbency and relaxation in facilitating sleep onset. When you or I say, "I am going to sleep," we usually mean, more precisely, that we've noticed that it's bedtime or that we're tired and that we're therefore going to try to fall asleep. Once in bed, we will often adopt a posture that we've found particularly favorable to sleep onset. Mine is, un-

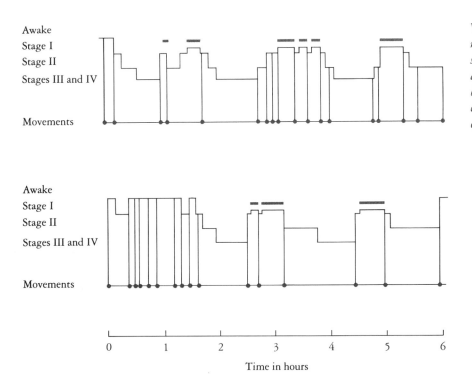

When the brain enters non-REM sleep, movement is suppressed. Thus body posture shifts can indicate prompt (above) and delayed (below) sleep onset. Once sleep begins, the long periods of immobility correspond to the recurrent non-REM episodes.

questionably, left side with left arm under pillow. I call this my sleep launch position. It almost always works. And I know that, if I make more than one or two posture shifts before falling asleep, I may be in for a bad night!

Muscle relaxation is one of two critical factors in falling asleep that we can voluntarily control. By reducing muscle tension, we reduce the stimulation to our brains that arrives via the muscle nerves. Since one of the main targets of these muscle signals is the brainstem, by relaxing we in effect help the brainstem neurons attain an excitatory level that is low enough to release other neurons with inhibitory influences necessary to induce sleep.

Another major source of excitation keeping the brainstem aroused is the cortex, seat of our ideas and our concerns. Its activity is also, to some extent, under voluntary control. Here again, we all know from experience that we can encourage sleep onset by letting go of our daily conscious concerns and substituting less relevant—or even nonsense—thoughts. This strategy is precisely what we have in mind when we say, "I think I'll go to bed and read a book." Detective stories do it for some of us, novels for others. And when all else fails, we can try counting sheep, imagining a pleasant scene, or humming a mystic mantra. All have the same goal and the same underlying mechanism.

IS SLEEP A RESPONSE?

Many scientists neglected the study of sleep in part because they had difficulty fitting this behavior into the stimulus-response formula that guided experimental research successfully in so many areas. Physiologists studying vision, audition, or balance were used to presenting an animal with a stimulus and tracking the response of the nervous system. Similarly, Webb's behaviorist predecessors paired rewarding stimuli with neutral stimuli and recorded responses (if they were Pavlovian), or they paired responses with rewarding stimuli and recorded the change in response frequency (if they were Skinnerian). Sleep did not fit easily into the stimulus-response paradigm from either of two classical angles: the physiologists could not trigger it in a reflex way by stimulating the brain; behaviorists inadvertently discovered that animals who became habituated to repeated stimuli often slept, but this finding was viewed more as a nuisance than as an opportunity.

Although it has some responsive aspects, sleep will never be adequately explained as a reflex response. Sleep is more an adaptive tendency not to respond when conditions are unpropitious. Thus, rather than sleep being caused by a reflex mechanism, many reflex systems are shut down by sleep.

The instinctual nature of sleep is suggested by the fact that it will ultimately overcome vigilance in even the most furtive animals. As this white tail deer naps in its browsing ground, it is luckily "shot" only by the cameraman, not a more aggressive predator. But even in life-threatening situations, sleep will finally have its way.

IS SLEEP AN INSTINCT?

Certain nineteenth-century naturalists, such as Edouard Claparède and Nicholas Vaschide of France, were impressed with the attributes that sleep shares with instinctual behavior. Sleep fit the concept of an instinct in that: (1) the behavior was stereotyped, since prone postures, closed eyes, relaxed muscle tone, and decreased sensitivity to stimulation were all seen in many species; (2) the stimulus conditions that favored sleep were consistent—darkness in sighted species, light in olfactory-guided species; (3) the adaptive advantage (energy conservation) was intuitively apparent; and, (4) most important of all, there appeared to be a *drive* to sleep—the longer the waking period had been, the more likely sleep was to occur.

For more contemporary theorists, like the Italian neurobiologists Giuseppe Moruzzi and Mauro Mancia, the stereotype of the EEG cycle of non-REM and REM phases added solidity to the view of sleep as an instinct and invited further modernization of the idea.

Moruzzi and Mancia borrowed their concept of instinct from the ethologists, who focused their study on the question of which environmental stimuli triggered instinctual behavior in animals. Ethologists viewed all behaviors as interactions between innate propensities (or drives) and environmental contexts (or releasers). For the ethologists, the most significant behaviors—such as feeding and mating—were run as fixed-action programs. When internal drive was high and the appropriate releaser was present, the animal would perform the instinctual behavior. These behaviors were called consummatory because of their all-or-none character (that is, once initiated, they tended to run to completion). The behavior that preceded the consummatory act was called appetitive: food seeking in the case of eating; courtship in the case of sex. Moruzzi and Mancia viewed the characteristic presleep behavior of animals (such as grooming, nest cleaning, and smoothing) as the appetitive behavior sequences, with REM sleep as the consummatory act. In their view, REM sleep was strongly analogous to the fixed action patterns of wake-state instinctual behaviors.

One of the most attractive features of the concept of sleep as an instinct is the ease with which it can accommodate the ethological notion of an innate releasing mechanism. Such a mechanism could be a built-in set of neurological commands that generate a given behavior, either in response to a condition in the environment or an internal drive or a combination of the two. It was Moruzzi's recognition that sleep was actively determined by a specific brain mechanism located in a specific brain region that prompted him to move in this unusual theoretical direction. In ethological terms, the growing strength of activity in the sleep center of the brain would account for the increasing probability for sleep to occur with prolonged waking. In the early stages of the activation, the mechanism would cause the appetitive phase, and the animal would look for a suitable place to sleep. We humans can generally take our place of sleep for granted, but consider the anxiety we may feel when we're traveling to a new place without a hotel reservation and night begins to fall.

The need for sleep does not increase regularly with time but, rather, fluctuates with a period length of about 24 hours. The reason for this is that, as you now know, there are at least two brain centers for sleep: the sleep demand mechanism in the brainstem interacts with the circadian mechanism in the hypothalamus.

BEHAVIORS THAT ENCOURAGE SLEEP

Almost every one has a set of tricks to abet sleep onset: a not-too-good book, a glass of warm milk, and the counting of sheep are common examples. But—over and above their suggestive power—do these nostrums really work? If they do, how do they trigger sleep mechanisms in the brain?

Eating

Claims have been made that warm milk, or the amino acid tryptophan, which occurs naturally in milk, has the effect of promoting sleep. Yet there is little evidence to support such claims. For every study showing some benefit, there are four showing none. In a well-controlled study of rats, Webb found that feeding (versus not feeding) had no significant effect on sleep.

Exercise

Many people swear that they sleep better at night if they have had at least moderate exercise during the daytime. Two studies using animal subjects have shown that, following exercise, there are dramatic increases both in attentional lapses and in deep, slow wave sleep. In 1968, I studied the effects on cats of moderate levels of treadmill exercise (4 hours at 1 mile per hour). Both because they had been working and because they had not been fed, the cats were quite hungry, once the exercise was over. However, to get a food reward, milk, they had to press a button on a box at 30-second intervals, a task requiring alertness and attention. Whereas the cats had performed well on this task prior to exercise, their efficiency was now measurably reduced, and they mistimed even the first few intervals. EEG recordings made during these attentional lapses showed slow waves.

Many of the cats simply gave up on the food and curled up to sleep. Compared to the control condition, in which the cats were simply placed on the treadmill to prevent them from sleeping or eating but did not exercise, the time taken to fall asleep was greatly reduced. In addition, the sleep following exercise lasted longer. Similar results were obtained by Junji Matsumoto of Tokushima, Japan, in studies with rats.

Sex

The rabbit is typical of those herbivorous mammals that feed by grazing, fidget to detect predators, run to escape them, and sleep very little. In laboratory studies conducted by Jean-Didier Vincent at the University of Bordeaux, rabbits almost never entered REM sleep. A dramatic exception was their sudden entry into REM following copulation. It is as if some hormonal influence combines with muscle and vascular relaxation in the aftermath of intercourse to trigger the rabbits' REM sleep mechanism. Since REM sleep is normally gated by the hypothalamus and the hypothalamus is the seat of sex hormone release, the link may be made in that structure.

We have at least anecdotal evidence that some such process may also occur in humans. Thus, it may not be entirely coincidental that so many human sexual encounters just precede sleep. And we might reasonably suppose that sexual orgasm has several sleep-promoting aspects: it not only leads to muscle relaxation, but also clears the cerebral circuits of tedious humdrum.

Moreover, the relationship between sleep and sex might be a two-way street. In keeping with previously noted facilitation of sexual arousal during REM sleep are reports by some people of enhanced sexual excitability at or even after sleep onset. Whether this is the result of diminished psychological inhibition or physiological weakening of cortical control is not known. The upshot is that sex may be a physiological sedative, and sleep may be a physiological aphrodisiac.

Familiarity

It may already be sufficiently clear that animals, including people, sleep best in familiar situations. Studies also show the beneficial effect of familiarity. For example, Webb found that the more rats had slept in a place, the better they were able to sleep there again. Nor is place of sleep the only condition that should be familiar. In a related study, Webb found that even after sleep deprivation, at which point recovery sleep is needed, rats performed stereo-

Animals sleep best in familiar settings, such as the warm, enclosing nest enjoyed by this dormouse. In this case familiarity breeds intimacy and security—not contempt!

typed behavioral rituals prior to going to sleep. This observation seems to indicate that the conditions most appropriate to sleep are sought even when that state is very strongly compelled by internal forces.

SLEEP DEPRIVATION

Practically everything that can happen to us, including all those things just listed as sleep enhancing, can interfere with sleep. Sleep is a fickle, highly variable state and one that is very easy to disturb. More interesting than an enumeration of the things that can disturb sleep are quantitative studies of what happens when sleep is intentionally disturbed.

Since the late nineteenth century, there have been a number of studies of total sleep deprivation in human subjects. Perhaps the most heroic is the American psychiatrist Anthony Kales' rigorously controlled study in the early 1970s, in which four human subjects were each deprived of sleep for 205 hours. While this and all other studies agree that sleepiness is a certain consequence of sleep deprivation, whether or not task performance suffers depends on the nature of the task. Even when performance of a task is affected, performance decrements are relatively limited. Humans have a great capacity to endure sleep deprivation and to compensate for the ensuing sleepiness.

Thus, while hallucinations and illusions do occur, they are quite rare before 60 hours of deprivation. Reflexes are generally unimpaired, although tremors, eye position instability, and reduced pain thresholds all denote functional neurological changes. Heart rate, blood pressure, respiration, and temperature measures show only minimal changes after up to 160 hours of deprivation.

In response to a pervasive and what seems to me perverse wish to get by on less sleep, researchers have studied the effects of limited sleep. A sleep ration of 3 hours per day was found to impair performance of tasks within a week. A ration of $5\frac{1}{2}$ hours was better tolerated but still impairing. And none of six couples who determined to lop another half hour off their daily sleep allowance each week reached the $4\frac{1}{2}$-hour mark before succumbing to sleepiness. One can only marvel at this evidence of man's experimental curiosity.

SELECTIVE DEPRIVATION

As soon as the dichotomy of sleep states was discovered, it was natural to wonder whether they had different functions, and, if so, whether deprivation of a specific state would have specific consequences. Because REM sleep was

In human sleep deprivation experiments, the number of awakenings needed to prevent REM sleep increases dramatically on successive nights for two subjects. The data for one subject is shown in red, and the data for the other subject is shown in beige. Each subject was observed over two extended periods, once for six days and once for ten days. On the tenth night the subject represented by red had to be aroused fifty times to interrupt attempts to enter REM.

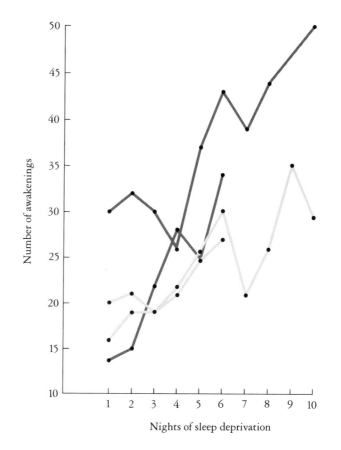

correlated with dreaming and because dreaming was considered to serve as a psychic escape valve, it was widely supposed that a loss of REM sleep would lead to the hallucinations and delusions of psychosis. And such a result was reported by the American physiologist William Dement in 1960.

Anthony Kales studied the effects of REM sleep deprivation in two graduate students, each of whom underwent a series of 6 and 10 successive nights in the laboratory. The results strongly and unequivocally confirmed a different finding of Dement's: more and more awakenings are required on successive nights to prevent REM sleep from occurring. The graph on this page shows that in the 6-night series, one subject who was awakened 10 times on the first night, had to be awakened 33 times on the last night. In the 10-night series, the other subject was awakened 30 times on the first night and 50 times on the last night. The important implication of these results is that there is a very real and pressing need, or drive, for humans to enter REM sleep. The longer we try to prevent REM sleep, the harder our brain tries to get it.

This drive also expresses itself in the amount of recovery sleep needed after the deprivation program is stopped. In the 6-night program, the subjects showed a REM sleep increase over preexperiment levels of 56 percent on the first night after the program and 42 percent on the second night. After 10 nights of deprivation, the corresponding figures were 90 percent and 80 percent. In absolute terms, this means that subjects had increased the amount of time spent in REM sleep, from 2 hours per night before deprivation to 4 hours per night after deprivation. These findings lead to the conclusion that the brain "knows" it needs to make up for lost REM and that it also knows how much it needs to make up. Particularly interesting is the fact that this rebound of lost REM persists for several weeks.

However, Kales failed to confirm Dement's psychological findings. Kales' subjects showed only minimal and insignificant changes in an extensive battery of psychological tests, even after 10 days of deprivation. In particular, they did not exhibit pathological personality traits, depressed mood, or cognitive impairment. Their irritability was attributable to loss of sleep generally, and not to loss of REM sleep per se. The failure of experiments to show that REM sleep deprivation causes specific deterioration of psychological functioning does not conclusively prove either that REM sleep has no psychological function or that it has no unique function. To be effective, deprivation may have to be continued for periods far beyond the limits morally allowable for human experimentation.

The duration and intensity of slow wave sleep, as with REM, increases steadily the more time has elapsed since it last occurred. But deprivation studies have been no more successful in assigning a special function to this stage of sleep than they have to REM. Based on these results, it is evidently easier to move Stage IV to the last third of the night than it is to move Stage I REM to the first third. This tendency may reflect the apparent priority given to Stage IV sleep both in evolutionary history and in each night of adult human sleep. Such a conclusion is supported by the observation that following total sleep deprivation Stage IV is recovered before Stage I REM.

SLEEP DEPRIVATION, THERMOREGULATORY DYSCONTROL, AND DEATH

Until recently, many sleep scientists were so discouraged by the lack of specific clues to the function of sleep that they began to accept the popular view that sleep was unwanted and useless, a kind of behavioral appendix that we should learn to do without. These fantasies quickly evaporated when Allan

Rechtschaffen reported, in 1983, that sleep deprivation, when well controlled and sustained, was uniformly fatal to rats.

The method Rechtschaffen used was simple, clear, and effective. The experimental setup is illustrated below. Electrodes that would record sleep and body temperature were attached to two rats. The rats were then placed on a revolving disc in a Plexiglas cylinder that had a Plexiglas partition, so that they were separated but visible to each other and to the experimenter. Both rats had free access to food and water. While any attempt to sleep on the part of the index rat caused the disc to rotate, the control rat was allowed to sleep ad lib. The procedure succeeded in reducing the amount of sleep in the index rat to 1 to 2 percent of that in the control animal.

No striking differences were observed in the first week to 10 days (the usual limit in most of the human experiments). Toward the end of the second

The control rat can sleep but the experimental rat cannot in this experimental setup designed by Allan Rechtschaffen. Because the sleep brain waves of the experimental animal automatically triggers movement of the cage floor, he wakes up after the briefest doze (violet), whereas the control animal is allowed to sleep peacefully (aqua). After several weeks the experimental animal begins to lose weight while eating more. Ultimately it fails to control body temperature and dies of overwhelming sepsis, evidence that immune functions have failed.

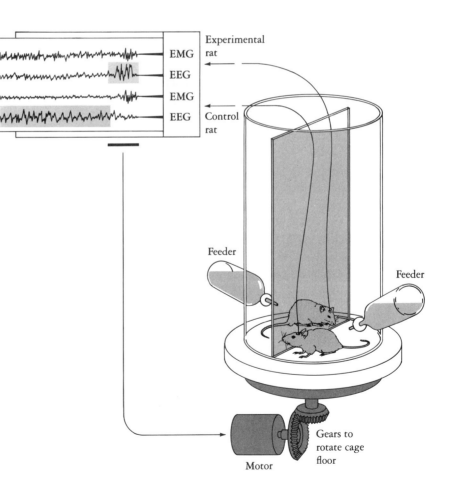

week, the sleep-deprived rat began to show signs of deterioration. For example, the skin of its paws became ulcerated. Even more striking, the animal began to lose weight despite eating more. The weight loss was not an effect of exercise, because the control animal did just as much muscular work but did not lose weight and did not increase food intake. It was not the result of starvation, since both rats had unlimited amounts of food. It was, rather, a state of metabolic dyscontrol, in which more and more calories were needed to restore energy balance. En route to the rat's ultimate demise at four weeks, weight plummeted while food intake soared. Furthermore, the capacity to regulate body temperature and immune function was lost.

What brain systems maintain the balance of our energy economy, and how does sleep help those systems to work? In Chapter 6 you will learn that when mammals sleep they may stockpile the brain neurotransmitters norepinephrine and serotonin, both of which have been implicated not only in thermoregulation, but also in such functions as attention and learning.

CULTURAL ASPECTS OF SLEEP BEHAVIOR

No discussion of sleep behavior would be complete without at least a brief look at relevant cultural factors and values. In modern, industrialized societies, these tend to conflict with and compromise biological realities. Our capitalist society by and large devalues sleep, viewing it as time taken away from productive and acquisitive pursuits. And we're not alone in this view. A major thrust of the ill-conceived Soviet sleep program was the development of an electrical brain stimulation device that would produce more efficient sleep. If Soviet workers only needed four hours of sleep, they could stay in the factory longer and the five-year plans might work!

The invention of the electric light has made the night an invadable temporal niche, and industries have moved into it. Shift work, like intercontinental travel, can cause us to want to be awake at times when our circadian clocks are chiming "sleep!" The urban areas in which so many of us live, with their all-night neon glitter and the incessant roar of traffic, add further insult to the injuries wrecked upon sleep. Add to this the anxiety caused by ambition, the sedentary life-style caused by bureaucratization, and the isolation of apartment house living, and you have an ideal formula for insomnia.

In view of the convincing evidence of sleep's behavioral significance, it behooves us to maximize the factors that we know to be sleep facilitating and to minimize those that we know to be sleep inhibiting. An essential first step is to assume an attitude more respectful of the value of sleep as essential to life.

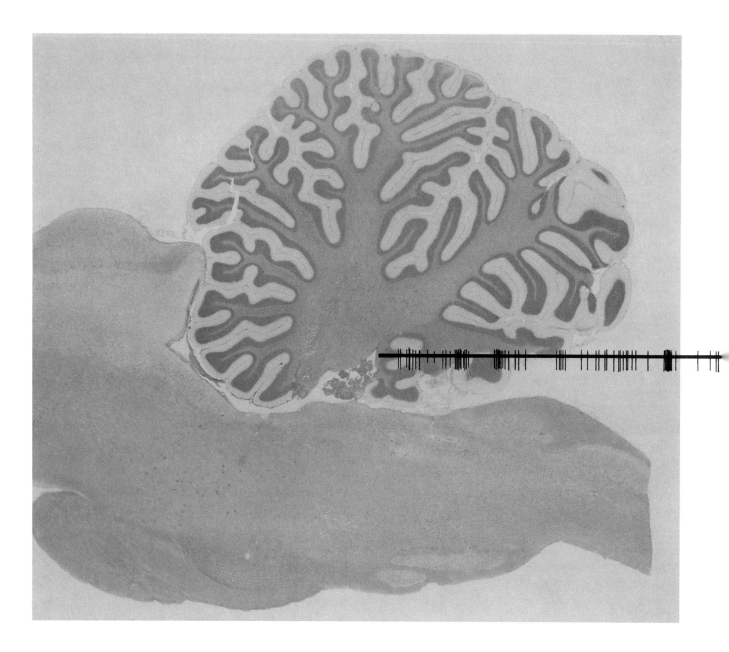

6

The Neurology of Sleep

Scientists know that the hypothalamus and brainstem control the states of waking, non-REM, and REM sleep. They would like also to understand how the switch is made from one state to another and how each state is maintained for a while and then terminated. Although humans and other animals take account of environmental conditions and can postpone sleep if those conditions are not favorable, the progression of states nonetheless is continuous and spontaneous. To change state so consistently, the brain must have built-in mechanisms that are something like clocks. Mechanical or electronic devices that change state regularly and rapidly are called oscillators, and this term has been used by neurobiologists who study the brain mechanisms that time behavior. Because the fundamental parts of such mechanisms are groups of neurons, scientists call such brain time-keeping mechanisms neural oscillators.

Recordings of individual cell activity have established the key role played by brainstem neurons in the generation of REM sleep events. The tracing illustrates the periodic firing of a pontine reticular neuron during REM sleep. The cell fires whenever an eye movement occurs.

THE CIRCADIAN CLOCK

The circadian clock, located in the suprachiasmatic nucleus of the hypothalamus, controls the sleep-wake cycle through its influence on the non-REM/REM clock in the pons. The means by which it exerts its influence are not as yet known. Perhaps the circadian clock conveys its message to the pons by sending neuronal signals arranged in a specific code. Or perhaps it sends discrete chemical molecules to the pons. It seems equally likely that the command signals to sleep or wake up are conveyed in a graded manner. If so, the non-REM/REM oscillators may read changes in the quantitative level of activation of the circadian system, rather than detecting some qualitatively distinct message.

If the signal does in fact take the form of variations in the level of activation, then a relatively small daily fluctuation in the spontaneous firing rate of one group of neurons in the suprachiasmatic nucleus could signal the end of sleep. The firing of these neurons could in turn be amplified by a sensory input, such as light. If the presence of light were to increase neuronal firing, then not only would the non-REM/REM oscillator receive an extra push at dawn to awaken us, but the time of awakening would be automatically adjusted to follow the changing seasons. Having been reset by light each morning, the circadian oscillator might then simply run down over the day, perhaps with the depletion of a chemical neurotransmitter. This "relaxation oscillator" model of the circadian clock has its advantages, but it does not successfully account for the relative independence of the timing mechanism from changes in body temperature.

The fact is that scientists know much too little about how the suprachiasmatic clock is organized—at either a cellular or a chemical level—to decide between many theoretically possible arrangements. Thus, although the Japanese physiologists Shin-Ichi Inoue and Hiroshi Kawamura have found that fluctuations in the *net* level of neuronal activity in the isolated suprachiasmatic nucleus follow a circadian cycle, they have not been able to tell us yet whether the clock consists of one group of cells firing at changing rates throughout the day or two groups of cells that fire in a reciprocal pattern. In the latter case, one group's activity would be at its peak when the other's was at its ebb. To detect such out-of-phase firing, scientists would have to record from individual cells, a procedure not yet possible because the neurons are too small to be resolved and recorded for many days using currently available microelectrode techniques. So scientists must wait for more precise data on this point.

THE RETICULAR FORMATION AND CHANGES OF STATE

The brainstem is clearly positioned and designed both to coordinate the activity of the spinal cord by switching various spinal reflex circuits in and out (as in respiration, chewing, swallowing, and walking) and to set the general level (or state) of the whole system, so that the organism is more or less likely to perform the various behaviors in its repertoire. With the addition of the more complex information-processing hardware of the forebrain, needed for analytic, abstract, and language functions, the brainstem continues its role as state coordinator by functionally linking the forebrain and spinal cord.

The early work of Moruzzi and Magoun demonstrated that electrical stimulation of the reticular formation of the brainstem had effects that ascended toward the cortex and effects that descended toward the spinal cord. When they stimulated the reticular formation of either the midbrain, thalamus, or hypothalamus, the animal became more alert and EEG activity increased (see the upper diagram on page 120). These effects in the cortex were associated with increases in both muscle tone and the excitability of spinal cord reflexes, as would be appropriate in organizing an animal to act against a predator, whether by defensive aggression, flight, or freezing.

Paradoxically, electrical stimulation of the reticular formation in the medulla sometimes produced the opposite effect: inhibition of spinal reflex activity (see the lower diagram on page 120). The resulting push-pull system of activation and inhibition playing down upon the spinal reflexes would have the advantage of allowing muscle tone to be continuously graded. But it was not until the discovery of REM sleep that the adaptive significance of the all-or-none action of this inhibitory apparatus was fully appreciated. In REM sleep, the upper motor systems of the cortex are turned on. But when the inhibitory apparatus of the medulla is spontaneously activated in REM, it mediates the postural atonia and preserves sleep by actively blocking motor output. Modern physiologists have now worked out many of the cellular details of spinal inhibition and shown us how it is switched on and off during the states of waking and sleep. Thus, the brainstem can arrange to have both the cortex and the spinal cord turned on (waking), both turned down (non-REM sleep), or the cortex turned on while the spinal cord is turned off (REM sleep).

FROM WAKE TO NON-REM SLEEP

Rarely do we find ourselves suddenly plunged into the depths of sleep from alert waking. We pass easily and imperceptibly from a high level of alertness to a low one. Our experience of gradual sleep onset fits nicely with a physiological

Top: *Electrical stimulation of the midbrain reticular formation excites neurons that project to the thalamus. The axons of cells in the thalamus then relay the excitation to the cortex. This cellular process is responsible for the spontaneous EEG activation of both waking and REM sleep.* Bottom: *Stimulation of the medullary reticular formation excites neurons whose axons inhibit the motor cells of the spinal cord. The resulting paralysis of the muscles occurs spontaneously in REM sleep but never in waking.*

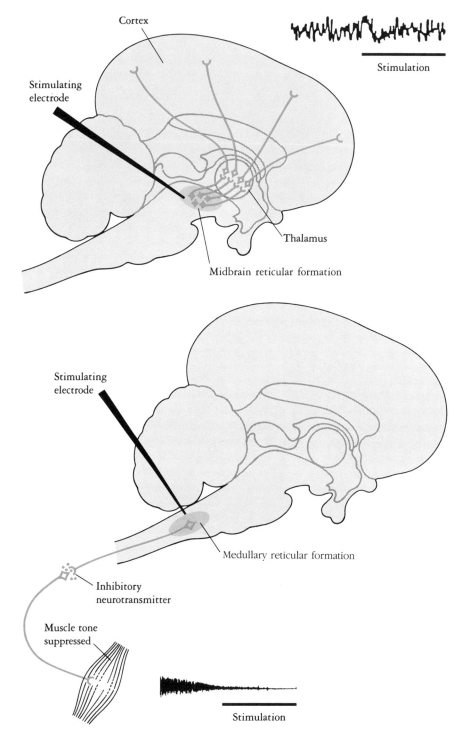

Cortex

Stimulation

Stimulating electrode

Thalamus

Midbrain reticular formation

Stimulating electrode

Medullary reticular formation

Inhibitory neurotransmitter

Muscle tone suppressed

Stimulation

fact of life: the alertness-mediating neurons in the midbrain, thalamus, and cerebral cortex slow down during sleep but do not abruptly cease firing. These millions of neurons change their output in a smoothly graded manner across states. Our graphic means of representing sleep as a staircase is thus a misleading artifact of analytic convenience; it helps us demarcate and enumerate phases of what is really a smooth, continuous process.

And yet, once the activity level of our arousal system has declined sufficiently—owing to both a diminished drive from the circadian clock and a fall in sensory input at night—we want to sleep definitively, not just to doze fitfully. We want to go from Stage I—which we can easily reach in classrooms and armchairs—through Stage II (with its spindles, sure signs of sleep)—down the sleep stage staircase and up again. While reading in bed, we let arousal level fall a bit until the pull of sleep is irresistibly strong; we then put out the light and assume a fully recumbent posture, thereby releasing our down-sliding brain and mind from their stimulus-monitoring tasks. No light, little sound, and no body posture signals to process. We hope this does the trick. But just what is that trick as far as our brains are concerned?

When nature discovers a good way of solving a problem, she uses the formula over and over again. Thus, we find yet another neural oscillator contributing to the process of sleep. At definitive sleep onset, neurons in our cortex and thalamus participate in an oscillation that is rather nicely depicted by the zzz's seen floating above the heads of sleeping comic book characters. Once the oscillation is great enough, brain wave spindles are produced and this part of the brain is no longer able to process external information well. The thalamus and cortex become susceptible to the normal loss of consciousness that we experience in sleep. To understand this mechanism, we need to know a little more brain anatomy and physiology.

During waking, the cerebral cortex receives sensory messages from the outside world via the thalamus. Each of the senses has its own sector of both thalamus and cortex. For example, the visual impulses from the retina go to the lateral geniculate nucleus of the thalamus and are relayed to the occipital cortex (the visual processing area of the cortex). There are, in addition, so-called nonspecific sectors of the thalamus, which also project to the cortex. But the pathways between the thalamus and cortex are not one-way streets. The neurons in each cortical area send connections back to the appropriate thalamic neurons, forming what is called a thalamocortical loop. It is as if each part of the system needs to know as precisely and instantaneously as possible what the other is doing. The illustration on the next page shows a simple diagram of the thalamocortical system.

The flow of information in these millions of thalamocortical loops is apparently kept orderly by means of several neuronal control systems. One is

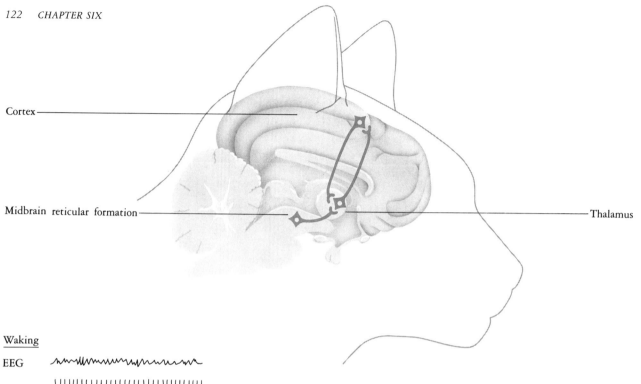

Cortex

Midbrain reticular formation

Thalamus

Waking

EEG

Neuron

Sleep

EEG

Neuron

Sensory information is relayed from the thalamus to the cortex, which in turn sends signals back to the thalamus, forming a thalamocortical loop. The activity of the loop is controlled by signals from the midbrain reticular formation. When the midbrain reticular neurons diminish their firing rate, thalamocortical and corticothalamic neurons begin unrestrained rhythmic firing. The result is EEG deactivation, leading to non-REM sleep and the EEG slow waves and spindles that characterize that state.

the midbrain reticular formation, whose neurons keep the minimum level of activity in the thalamocortical loops during waking high enough so that external data can be processed. If the level of activity is kept high, the neurons in the circuit tend to fire more regularly, and any incoming signal will be more easily discriminated because it produces an irregularity in the ongoing signal processing. When listening for footsteps in a house at night we freeze, reducing self-generated inputs and turning up our thalamocortical circuits to pick up the faintest perturbation.

Because of the strong positive feedback that arises from the two-way connections, the thalamocortical loops have a tendency to begin oscillating. When this happens, the cortical neurons begin to fire rhythmically in two-step time with their partners in the thalamus. As a result, the signals arriving at the thalamus on sensory channels cannot be discriminated from intrinsic noise. This part of the brain can no longer process sensory data, and loss of consciousness results.

Without controls, the brain would be liable to go into spontaneous oscillation every time a stimulus impinged upon it. When uncontrolled and pathologically exaggerated, the oscillation becomes an epileptic seizure, with its brief lapses of attention (so-called petit mal) or major losses of consciousness (so-called grand mal). Such pathological oscillation is prevented mainly through inhibition, which is of at least two types: local and remote. Both local and remote inhibitory neurons are called interneurons because they are connected in parallel with the cells whose output they control. Interneurons provide local control in both the cortex and thalamus by immediately converting the excitatory signals generated by the thalamocortical neurons into regulatory inhibitory signals. Via this prompt and precise negative feedback, the thalamocortical system is further tuned and balanced. In other words, the harder it is driven by inputs from any source, the more it is restrained from spasmodic firing.

Remote inhibitory control is exerted when the corticothalamic loops excite the nucleus reticularis thalami, a specialized shell of neurons wrapped around the thalamus, into which the reticularis thalami projects its inhibitory fibers. This remote inhibitory system may provide a kind of sector control that enables the system to switch from one stimulus mode to another, for example, from the visual to the auditory mode. The reticularis thalami is a kind of channel selector, as it were.

Whatever the precise function of these fascinating thalamocortical circuits may be during waking, the circuits change dramatically in non-REM sleep. Many neuroscientists reasonably suppose the thalamocortical loops to be the neuronal basis of our loss of attention, our loss of external processing capability, and our conscious oblivion upon falling asleep. At sleep onset, the brainstem excitatory drive upon the thalamocortical system falls, and as it falls, so does the level of activity in the thalamocortical loops. When activity drops to a critical level, the threshold for spontaneous reverberation of these looplike circuits is passed, and they begin to oscillate. Here we see again the concept of the brain as a set of coupled oscillators. The detailed sequence of events has been worked out by the Rumanian-born neurophysiologist Mircea Steriade, now working in Quebec, Canada.

With each reverberation or oscillation, more and more of both the thalamic and cortical neurons are caused to fire. The successive waves recorded by the EEG augment in amplitude, reach a peak, and then decrease as the inhibitory interneurons gradually impose more restraint. It is this stepwise increase and decrease in wave amplitude that gives its distinctive form to the EEG spindle, which tends to recur every 10 seconds or so in Stage II of non-REM sleep.

Little is known about the exact mechanism by which non-REM sleep begins. Because of the clear relationship of sleep to the circadian clock, it is reasonable to suppose some interplay between neurons in the hypothalamus

Left: *During waking, thalamic neurons fire regularly in response to steady excitation from the midbrain reticular formation.* Right: *During sleep, the reticular excitation decreases, and the thalamic and cortical neurons begin firing in alternating intermittent bursts. The intermittent pattern of firing gives rise to spindles in the cortical EEG. Steriade has recently shown the cholinergic neurons of the brainstem play an important role in preventing this thalamocortical ascillation during both waking and REM sleep.*

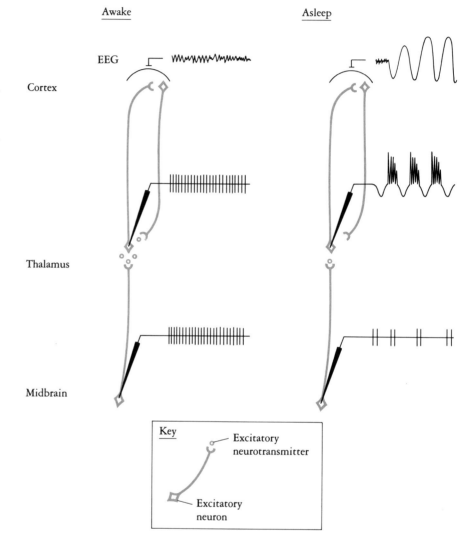

Key

Excitatory neurotransmitter

Excitatory neuron

and lower brainstem. The MIT neurobiologist Walle Nauta has shown that animals became hypervigilant when an area next to the hypothalamus called the basal forebrain is destroyed. Scientists working at UCLA have induced slow wave sleep by stimulating this same region. However, single-cell studies have not yet revealed neurons with the strong firing prior to sleep onset that a theory of active sleep control by this region predicts.

In closely related work, the American physiologist John Pappenheimer has demonstrated that Factor S, a peptide molecule extracted from the fluids of sleep-deprived animals, can induce slow wave sleep when injected into the hypothalamic area of the brain (but not the brainstem). Factor S is also strongly sleep promoting when injected into body cavities. It has an interesting tendency to raise the animal's temperature as if it played upon thermoregulatory control as well as sleep-generating mechanisms. The peptide is known to play a role in triggering of Interleukin I, a protein that is synthesized by the body as part of the immune response. Pappenheimer's student, James Krueger, has developed a strong experimental program to explore the positive interaction between sleep and the immune system.

A multidisciplinary study of the basal forebrain may enable scientists to establish firmly the missing link between the circadian clock, energy regulation, and the mechanisms that control slow wave sleep.

THE BRAIN IN REM SLEEP

After we humans have been asleep for about an hour, the midbrain reticular formation becomes spontaneously reactivated, and with the gradual rise in its firing level, the tendency for thalamocortical circuits to reverberate is suppressed. The EEG is now desynchronized. We are entering REM sleep.

REM sleep alters the motor and sensory systems in a distinctive way. It turns on central motor neurons (producing the REMs and twitches) and, at the same time, turns off the motor neurons in the spinal cord that effect movement of other parts of the body, producing the paralysis we call atonia. In addition, the brainstem activation during REM sleep provides a powerful source of internal stimulation, the PGO waves, while simultaneously blocking external sensory stimuli. The result is an extraordinary set of paradoxes, of which our dreams are the conscious experience: we see things, but the lights are out; we imagine running, flying, or dancing the tango but are paralyzed; we explain the bizarre proceedings to our full satisfaction, but the logic by which we do so is as bizarre as the proceedings; we have intense emotional involvement in the action, but we forget the whole business as soon as it is over. What is going on? In each case, the answer is the same: something is added to and something else subtracted from the brain's wake state repertoire. The diagrams on pages 126 and 127 summarize the brain mechanisms that are special to REM sleep.

Sensory systems

The activity of individual neurons in REM sleep was first studied by Edward Evarts, who recorded from single cells in the visual cortex of cats. Evarts was surprised to find that these central mediators of vision were generally as active in REM sleep as during waking and that the neurons fired in clusters—just as they do when visually stimulated—in association with the REMs. Such activity could stimulate dream vision. Since that first report, many other sensory systems have been shown to be activated in a pulsatile manner in REM sleep,

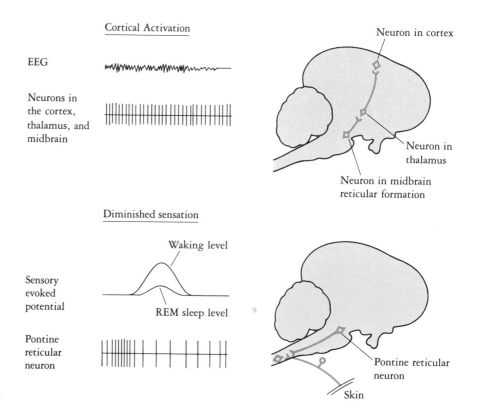

Many brainstem neuronal systems are orchestrated to produce REM sleep. Cortical activation is the result of high levels of regular discharge by neurons in the midbrain, thalamus, and cortex. We do not sense external stimuli very well because when pontine neurons fire in bursts, they inhibit nerve terminals from the skin at their endings in the spinal cord. The rapid eye movements are also a consequence of the burst discharge of pontine reticular cells, which directly excite nearby oculomotor neurons in the brainstem. We experience motor paralysis because reticular cells in the medulla, which fire at a high regular rate throughout REM sleep, directly inhibit the spinal motor neurons. PGO burst cells fire clusters of spikes prior to each PGO wave, signalling the direction of the REMs to neurons in the thalamus and cortex.

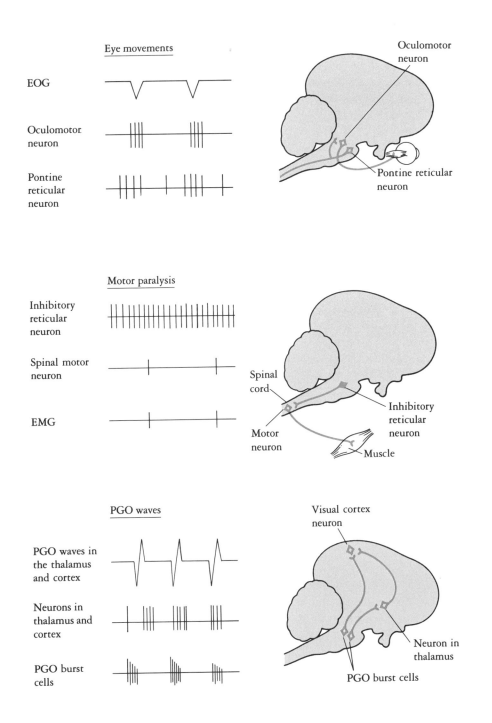

Eye movements

EOG

Oculomotor
neuron

Pontine
reticular
neuron

Oculomotor
neuron

Pontine reticular
neuron

Motor paralysis

Inhibitory
reticular
neuron

Spinal motor
neuron

EMG

Spinal
cord

Motor
neuron

Inhibitory
reticular
neuron

Muscle

PGO waves

PGO waves in
the thalamus
and cortex

Neurons in
thalamus and
cortex

PGO burst
cells

Visual cortex
neuron

Neuron in
thalamus

PGO burst cells

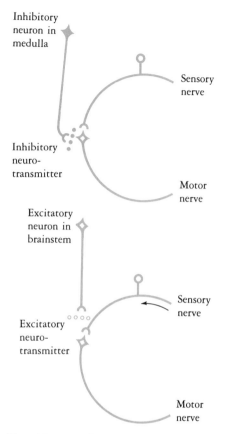

Top: *Motor paralysis in REM sleep results from postsynaptic inhibition of motor neurons. Inhibitory impulses arise from the spinal endings of neurons whose cell bodies are in the medullary reticular formation. These impulses decrease the excitability of the spinal motor neurons on which they impinge. When an impulse arises in a sensory nerve, it therefore has more difficulty exciting a motor response.* Bottom: *The diminished sensation of REM sleep is caused by presynaptic inhibition of the sensory nerve terminals. Excitatory signals from the reticular formation cause neurotransmitters to be released from the sensory nerve endings. When an impulse arises in a sensory nerve, its effect is weak because its terminals are depleted of excitatory transmitter.*

including the thalamic nuclei that relay information for audition, vision, and touch to the cortex, and including the brainstem vestibular system and cerebellum, mediators of body position sense. All these sensory modalities are prominently represented in dreams, as you will see in Chapter 7.

Given the increased excitability of the central pathways of sensation, it is all the more surprising that we don't wake up in REM sleep. To try to explain this paradox, the Italian physiologist Ottavio Pompeiano tested the excitability of the nerve endings from skin and muscle to the spinal cord. He found that they conducted external stimuli much less well in REM sleep than in non-REM sleep. The reason was found to be a process called presynaptic inhibition, in which a synapse becomes less effective owing to the depletion of its neurotransmitter by internal signals descending from the brainstem. Because the neurons with endings in our skin lack sufficient neurotransmitter, they are less able to transmit external signals to the brain. Thus, many noises, lights, and other interruptions of the night do not arouse us from REM sleep.

Motor systems

Following his study of the visual cortex, Evarts turned his attention to the motor cortex. There he was able not only to record from individual cells but from those very neurons that sent movement commands directly to the spinal cord. Again he observed high levels of clustered neuronal firing in REM sleep, and again the clusters appeared in conjunction with the runs of rapid eye movement. The high levels of firing were matched in waking only when the animal actually moved. This finding has been confirmed and extended to other motor systems, such as the red nucleus in the midbrain and certain cells of the pontine reticular formation, which are known to directly command eye movement. Why, then, don't we move during REM sleep? What blocks the central motor commands?

Pompeiano again provided the answer: inhibition. But this time his experiments indicated a postsynaptic process: the motor neurons of the muscles were less responsive to excitation by the motor cells from the cortex. The American neurophysiologists Michael Chase and Lloyd Glenn have confirmed Pompeiano's hypothesis by recording inside the motor neurons themselves. They measured the inhibition directly and found that it was strong enough to markedly reduce the probability of firing. The clear implication is that even though we might like to run away from our dream pursuers, we can't!

The idea that dreamed movements would actually occur if not so blocked is supported by experiment. After making small lesions below the locus coeruleus in cats, both Michel Jouvet and François Delorme (in France) and Kristin Henley and Adrian Morrison (in the United States) observed REM sleep with an astonishing panoply of organized—but entirely automatic—motor behavior, such as running and walking movements.

So now we can more easily understand our experience of movement in dreams. It is compounded of our sensory system's awareness of the motor commands and the related activation of body position sensors in the vestibular system. We can also explain a curious and striking but little-noted aspect of dreamed movement: most of it is effortless. Although we preserve the fiction that it is we who are moving—as if we willed it—we are, as much, moved—quite involuntarily—by our activated motor systems. In fact, when "we" become the active agents of dream movement, we may find it quite frustrating—the system won't respond quite right. When chased by dream pursuers, I often find my legs quite leaden and unwilling to follow my urgent appeals to carry me faster out of harm's way. This standoff between the now voluntarily commanded movement (saying go) and the involuntarily clamped muscles (saying stop) often causes the dream state to fracture, and we wake up with vivid recall.

Orientation and position control

During waking, eye movement, head position, and body position are coordinated by the brainstem reticular formation, the nearby vestibular nucleus, and, of course, the neurons of the visual cortex and motor cortex themselves. All three groups of neurons are powerfully activated in REM sleep, when they come to fire in rhythmic bursts. After having gathered sufficient intensity, these bursts will drive the REMs. In the pontine reticular formation, many neurons are actually more excited during REM sleep, than during waking.

So now I add a third factor to the REM sleep system: Not only are both central sensory and motor centers activated, but the brain's orientation and position control centers are hyperaroused. The brain is actively generating "false" data about the body's posture and location in space. No "true" data is available from the outside world because the sensory input channels are blocked, and no motor output can change body position. Thus, we experience both intense orientational instability and a vivid sense of body position change during dreaming.

Eye movement commands as sensory signals

As I have discussed on page 125, during REM sleep the brainstem sends spontaneously generated signals, recorded as PGO waves, to the thalamus and cortex. These waves have recently been discovered to contain information about eye movement. For example, if the eyes move to the right in REM

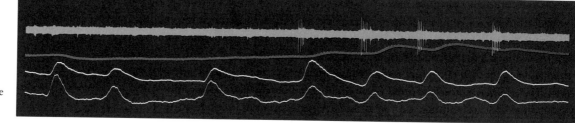

PGO burst cell

Eye movement

Left geniculate

Right geniculate

The tracings above record the activity of the PGO system during REM sleep. Specialized neurons in the pons fire a burst of action potentials (upper trace) whenever the eyes move toward one side (second trace). (When the eyes move to the left, the eye movement tracing moves upward, as in this case. A downward movement of the trace would indicate eye movement to the right.) At the same time that the eyes move left, there is a PGO wave in the geniculate body that is larger on the left side of the brain (third trace) than on the right (fourth trace). Thus, the PGO waves inform the sensory visual brain in the cortex that eye movements are occurring. The cortex can then construct a change of scene in the dream that corresponds to the eye movement.

sleep, the PGO wave is larger in both the right lateral geniculate nucleus and the right visual cortex. In other words, not only are the visual sensory areas of the brain activated, but they are bombarded with internally generated stimuli describing eye position changes. Because the PGO waves reach the thalamus and cortex before the eye actually begins to move, we can conclude that the visual brain activity anticipates changes in scene. Single-cell recordings have identified a population of neurons in the brainstem that appears to be responsible for sending these signals to the forebrain. Called PGO burst cells, these neurons are far more excitable in REM sleep than in waking.

Neurologists working with brain-damaged patients have suspected the existence of such sensory stimuli encoding movement commands. Patients who could not move their eyes owing to brainstem disease would experience a scene shift when they tried to do so. Their experience suggests how the brain's own bookkeeping system works: whenever an order to move is sent to the muscles, a memo is simultaneously sent to those receiving areas of the brain that were likely to be affected by the movement. The PGO waves reflect the automatic activation of this internal communication system in REM sleep.

Our dream plot thus thickens. We have found at least one internal stimulus source for our nocturnal visions.

A BRAIN CHEMICAL DEBIT IN REM SLEEP

The common-sense view of sleep shared by Pavlov and by Sherrington was that neuronal activity would slow down or even cease, explaining the loss of consciousness and the restfulness of sleep. But this view was immediately upset when Evarts found that most brain neurons slowed only slightly in non-REM sleep and then resumed wake-state levels of activity in REM sleep. Scientists quickly became so thoroughly adapted to this new REM-sleep activation rule that they had trouble recognizing a most intriguing exception to it.

Microelectrode recording is like trout fishing in that one spends hours getting the equipment ready, then waits patiently for some action and often comes home empty-handed. One day, the Canadian ophthalmologist Peter Wyzinski and I were fishing for big reticular neurons in the deep brainstem pool when we hooked a little fellow in the locus coeruleus that behaved so oddly we decided not to throw him back. This cell fired with monotonous regularity during quiet waking (when the big reticular neurons tend to be silent). It then slowed down at sleep onset, which is not, in itself, particularly noteworthy. But when it fired less and less and less frequently—and finally stopped altogether—we could not, at first, believe it. We thought the cell had gotten away: that it continued firing, but at a distance too great for our

Top: *The pontine brainstem is stained with a combined Nissl stain (coloring the cell bodies purple) and Klüver-Barrera stain (coloring fibers blue). Bottom: The small locus coeruleus cells which stop firing in REM sleep and the large pontine neurons which fire more are seen in these two blow-ups. The one on the left corresponds to the upper box, and the one on the right to the lower box in the top figure.*

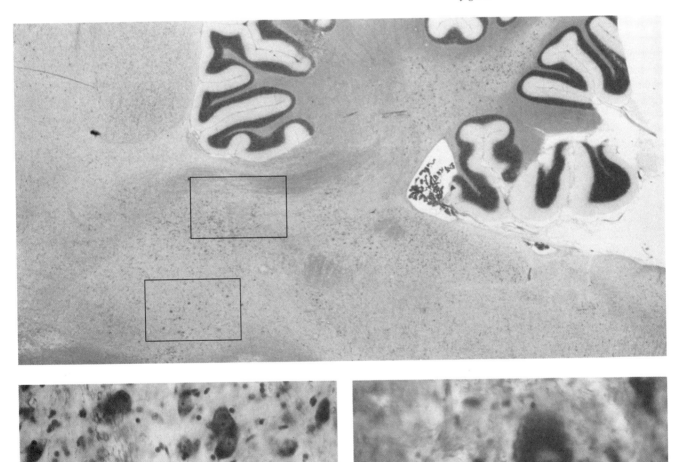

electrode to record it. But good trout fishermen that we were, we thought we might still have something, so we didn't move a muscle until the end of the REM period when, lo and behold, the cell recommenced its metronome-like wake-state rhythm of firing.

We now know that these locus coeruleus REM-off cells share their distinctive wake-on, REM-off pattern with neurons in the raphe nuclei and others in the peribrachial region of the pons. All such cells almost certainly release either norepinephrine or serotonin, and all of the cells that are likely to be of these chemically distinctive classes turn off in REM sleep. Since these neurons distribute their axons widely throughout the cortex, brainstem, and spinal cord, we can infer that although electrically activated, the brain during REM sleep is deficient in both norepinephrine and serotonin. Measurements of the concentration of these transmitters confirm this hypothesis.

Since these two neurotransmitters which are subtracted from the brain in REM sleep play strong roles in all leading theories of dreaming and memory, the implications of the findings for dream theory are far-reaching. They may solve the puzzles posed by dream bizarreness, by loss of self-reflective awareness, by disorientation, by logical fallacy, and—above all—by memory failure. These deficits are not simply due to the absence of external data—something has been taken away from the brain itself.

DISINHIBITION AND REM SLEEP EXCITATION

I have tentatively ascribed sleep onset to a gradual decrease in the excitability of the reticular formation, which is set off by a signal from the circadian oscillator. In so doing, I have left unexplained an important finding of Steriade's: the reticular formation is periodically reactivated and deactivated throughout the sleep period. I will now consider the possibility that the cause of the reticular activation during REM sleep is quite different from the cause of its activation during waking. To explain this process, I will introduce the paradoxical concept of disinhibition. Put very simply, a nerve cell will fire faster if something reduces the inhibitory restraint that normally holds its excitability in check. It's like taking the governor off the accelerator of an automobile.

The neuromodulators norepinephrine and serotonin act over a longer time and over a wider area than classical neurotransmitters. These characteristics imply that the two chemicals may serve to set up the proper background conditions for the brain to process information. According to one theory, the aminergic system allows better discrimination of the signals conducted by

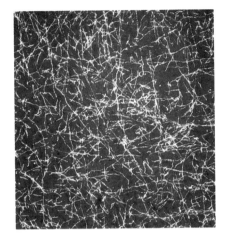

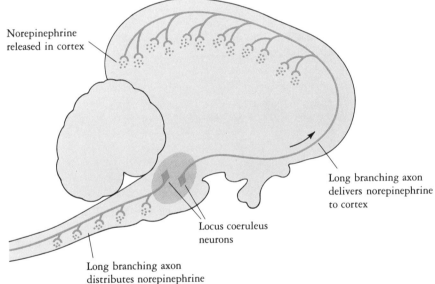

Norepinephrine
released in cortex

Long branching axon
delivers norepinephrine
to cortex

Locus coeruleus
neurons

Long branching axon
distributes norepinephrine
to spinal chord

During waking the neurons of the locus coeruleus distribute norepinephrine in large amounts to a widespread area of the cortex and spinal cord (center diagram). They secrete much less neurotransmitter in non-REM sleep and almost none in REM sleep. Norepinephrine-containing neurons in the locus coeruleus (bottom photo) and the dense meshwork of their fibers in the cortex (top photo) are visualized by the histofluorescent dye staining techniques of the neuroanatomist Reinhardt Grzanna.

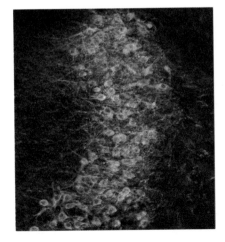

faster and spatially more precise sensorimotor circuits. The loss of aminergic neurotransmitters could be yet another reason for the brain's processing of sensory information to shut off during sleep.

Whether this view is correct or not, the neuromodulators *do* exert a predominantly inhibitory influence on other neurons. Thus, when the activity of the aminergic system begins to decline at sleep onset the brain begins to be progressively disinhibited as well as becoming suddenly disfacilitated. The result is an initial fall (owing to diminished excitation) and subsequent rise (owing to disinhibition) in reticular activation level. The initial fall is responsible for non-REM sleep events (such as the EEG spindles), and the subsequent rise is responsible for REM sleep events (such as EEG activation).

Aminergic deactivation and reticular activation are thus mirror image processes at all phases of the sleep cycle. The contrast between the two becomes particularly strong at REM sleep onset and REM sleep offset. This suggests that a reciprocal interaction between the neurons may be what turns REM sleep on and off.

THE RECIPROCAL INTERACTION MODEL

Now I consider the possibility that the key to the organization of the non-REM/REM sleep cycle may lie in the link between two populations of neurons in the pons. The fundamental idea is quite simple: the two populations form a push-pull, on-off neural oscillator. A simple diagram of the reciprocal interaction model appears on the next page.

According to this model, the REM-off cells in the aminergic nuclei set the level of excitability of the REM-on cells in the reticular formation (and elsewhere) via the mechanism of disinhibition just discussed. The chemical nature of the REM-on cells is unknown, but some of them may be cholinergic (they use acetylcholine as their neurotransmitter), and many of them may be cholinoceptive (they respond in an excitatory manner to acetylcholine released by other cells). The REM-on cells contact each other so that when one starts to fire others are likely to follow suit, and soon the whole population becomes active. This process is called self-activation.

Both groups influence each other and do so in equal but opposite ways. When the REM-off cells themselves are on in waking, they hold the REM-on cells in check. The REM-off cells are also self-restraining, since whenever they fire, they send each other a message not to fire for a while. This mechanism explains their low, metronome-like rates of discharge. In this way, the REM-off cells maintain a steady level of output over a long period of time in an efficient way. During waking our brains are thus suffused with a relatively uniform concentration of these molecules, just what is needed to maintain a steady state of arousal.

The REM-on cells have quite different properties. Owing to their self-excitatory feedback, they tend to fire *more* once they receive a strong enough excitatory signal. They thus discharge in intense clusters, with intervening silences when there is no excitatory stimulus. Firing in isolated bursts is *not a* good way to maintain a state, but it is a good way to get something done in

Reciprocal interaction between REM-on cells and REM-off cells is conceived to be the neuronal basis of the non-REM/REM sleep cycle. The individual cell recordings shown below the brain indicate how the firing of the two groups of cells changes from waking through non-REM sleep to REM sleep. Because REM-off cells are inhibitory, when they are active (in waking) they suppress the REM-on cells. As the activity of the REM-off cells declines (in non-REM sleep), that of the REM-on cells gradually increases. When the inhibition from the REM-off cells is at its lowest point, the excitatory action of the REM-on cells reaches its highest point, and a REM sleep episode occurs. For simplicity the REM-off cells are shown projecting only to the cortex and the REM-on cells only to the spinal cord, whereas in reality cells of both types project to both regions.

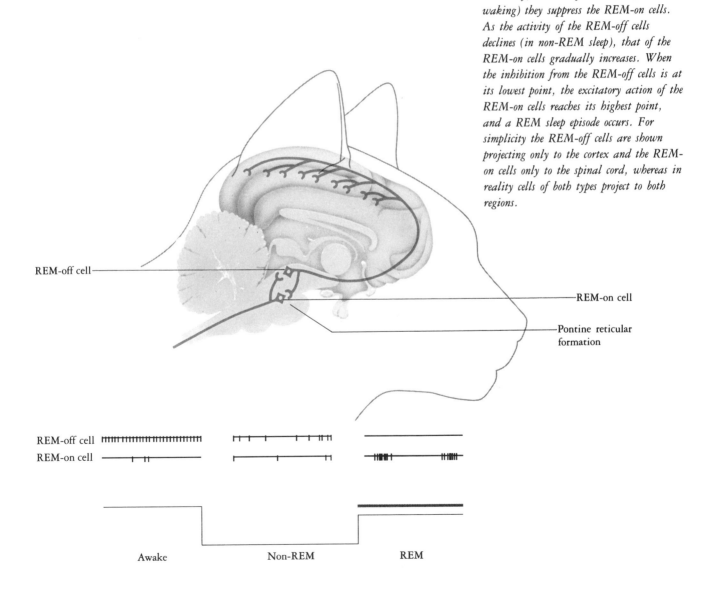

REM-off cell

REM-on cell

Pontine reticular formation

REM-off cell

REM-on cell

Awake   Non-REM   REM

One group of brainstem neurons that may supply acetylcholine to the REM sleep generator network of the pons is found in the pedunculopontine nucleus. These cells are assumed to be cholinergic because they contain an enzyme that is essential to the biosynthesis of acetylcholine. The stain used by the American neurobiologist Marcel Mesulam interacts with that enzyme, called choline acetyltransferase.

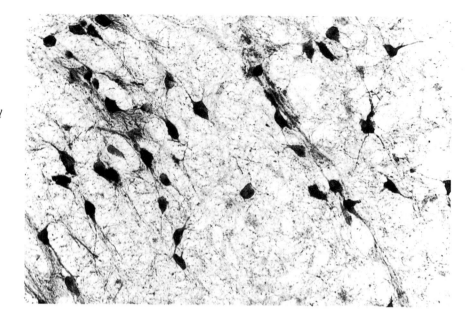

it—like move an eye (the REMs) or convey messages to other parts of the system (the PGO waves).

The work of Kazuya Sakai in Lyon, France, has revealed that there are two groups of REM-on cells that fire regularly and therefore *are* well-adapted to state maintenance. But these cells differ from REM-off cells in having much higher rates of discharge. One group (in the pons) projects directly to another group (in the medulla). Together they mediate the inhibition of spinal reflexes and the REM sleep atonia that prevent us from acting on the motor commands issued by the upper brain as we dream.

The virtual arrest of firing by aminergic neurons in REM sleep and the reactivation of cholinergic neurons back to or even above their waking state levels may profoundly affect the functioning of the brain. As a consequence of the changes in firing levels, the *ratio* or mix of the neurotransmitters would be markedly different in REM sleep from the ratio during waking. Thus, although both waking and REM are electrically activated brain states, the neurons of the brain would be exposed to completely different biochemical conditions in each state. The difference in neurotransmitter ratio might alter such functions as information processing and conscious awareness. The way that this ratio could change is illustrated in the graph on the next page.

Robert McCarley recognized that the physiological model he and I had sketched to explain the non-REM/REM sleep cycle was formally identical to that describing the interaction of prey and predator populations in field biology. In the ecological model, the growth of the predator group occurs at the expense of the prey. The growth of the predator population is slow at first because it takes time to reproduce no matter how abundant the prey. Growth then expands explosively, but slows and ends as the supply of prey is exhausted. The predators go into decline and their numbers remain small until the prey population recovers. The entire process can then repeat itself. Because of the delays caused by the time needed by both groups to reproduce, the two curves representing the numbers in each group are out of phase with each other (see the graph on page 47).

In the neuronal model, McCarley lets the REM-off cells be represented by the predators and equates predation with their suppression of REM-on cell excitability. And he lets the number of individuals in the ecological model represent the mean excitability level of the two groups of neurons. The analogy of prey exhaustion, leading to the decline of the predator group, is exhaustion of inhibitory neurotransmitter, which allows the level of activity to increase.

The prey analogue is the REM-on population, whose excitability grows exponentially as the predatory inhibition declines. By recruiting more and more cells, the state of the brain and mind is changed to REM sleep-dreaming. After rising to its peak, the excitability of this system gradually declines, both because its own excitatory neurotransmitter becomes exhausted and because the inhibitory power of the REM-off group gradually recovers. When this balance shifts radically, the REM period ends and the cycle recommences.

The practical value of the mathematical model is predictive. For example, because the two curves are slightly out of phase, McCarley and I were able to notice that the REM-off populations did anticipate—and we suppose precipitate—the end of REM sleep by gradually resuming firing in the last portion of each REM-sleep epoch.

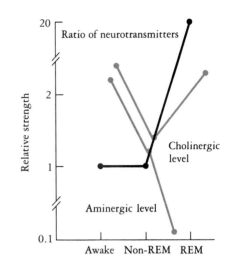

If the release of norepinephrine and serotonin decreases in parallel with the decreased discharge rate (orange), while the release of acetylcholine increases (purple), the relative strengths of the two kinds of neurotransmitter would change as shown in the graph above. As a result, the ratio, C/A, of the concentration of the two classes of neurotransmitter will be much higher in REM sleep than in the other two states (black). The brain is thus exposed to very different biochemical conditions in REM sleep.

THE EXPERIMENTAL CONTROL OF REM SLEEP

My colleagues and I wanted a way to attack the question of neurotransmitter control of REM sleep. For example, were the REM-on cells really excited by acetylcholine? If so, then injecting drugs that enhanced or imitated acetylcholine should cause REM sleep to occur earlier and for longer times than under control conditions. We were able to test these theories by using a reliable and accurate system for delivering known quantities of drugs to precisely localized sites in the brainstem of awake animals.

The first test of the reciprocal interaction model was to try to enhance REM sleep by bolstering the positive feedback in the REM-on cell population. Two graduate students, Thomas Amatruda and Thomas McKenna attempted

In order to learn exactly which cells are affected by those cholingergic drugs that trigger REM sleep, Harvard neuroscientist James Quattrochi attached fluorescent beads to carbachol (a drug resembling acetylcholine) and injected it into the brainstem. Not only does the injection site light up (left), but all the cells that project to it also fluoresce (right). This occurs because the beads are taken up by the axons and pumped back to cell bodies, a process called retrograde transport.

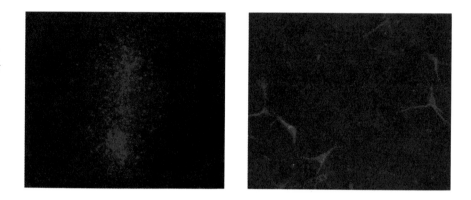

to increase feedback by injecting drugs that looked to the brain like acetylcholine. In these most dramatic experiments, they saw cats who had been wide awake enter REM sleep in three minutes and stay there for $2\frac{1}{2}$ hours!

The neurobiologist Helen Baghdoyan found that only injections in the pons enhanced REM sleep; injections in the midbrain and medulla produced arousal and suppressed REM sleep. Within the pons, the precise optimal site of injection was in the reticular formation just behind the noradrenergic locus coeruleus. We still don't know precisely which—and how many—brainstem neurons need to be activated to change the state of the entire brain and mind. We think the number must be quite small because limiting the spread of the injected drug within the pons does not diminish the effect. But we *do* know that the same effects can be achieved by increasing the amount of—rather than imitating—the brain's own acetylcholine. By blocking the enzyme acetylcholinesterase, which normally breaks down the acetylcholine molecules as soon as they are released, the concentration of the naturally occurring transmitters gradually builds up. The high concentration of acetylcholine triggers spectacularly long and intense artificial REM periods.

The data from drug studies shows that REM sleep injection sites receive input from both aminergic and cholinergic sources. Every neuron that sends an axon to a REM sleep injection site has been identified and entered into the appropriate place in these three-dimensional computer maps of the brain. Left: Reticular, REM-on neurons are blue, and aminergic, REM-off neurons are red (dorsal raphe) and yellow (locus coeruleus). Right: Three possibly cholinergic populations project to the injection site: the dorsolateral tegmental nucleus (white), the peribrachial nucleus (pink), and the pedunculopontine nucleus (red-orange).

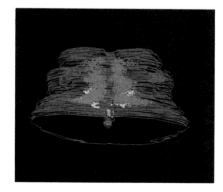

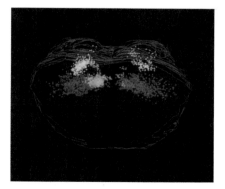

Although the direct introduction of drugs into the human brain is impossible, cholinergic systems of the brain *can* be enhanced by the intravenous delivery of compounds, which then cross into the brain. In humans, as in the cat, the most effective drugs are those that mimic the action of acetylcholine and are hence called cholinergic agonists. When the cholinergic agonists pilocarpine and arecoline were infused into sleeping people by Christian Gillin and his team at the National Institute of Health in Washington, the latency (or time delay) to the first REM period was shortened, and that REM period was lengthened. The subjects reported dreaming when awakened. Some of these results for both cats and humans are summarized in the diagram on this page.

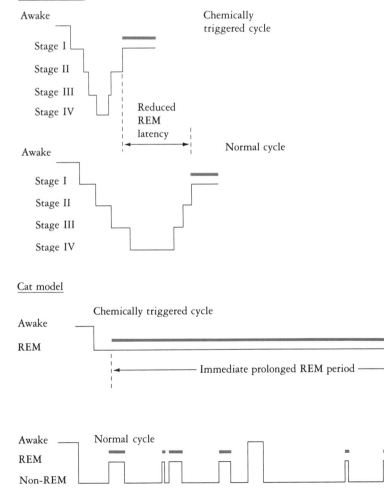

By stimulating the brain with cholinergic drugs, we can enhance REM sleep in humans and in cats. When humans are injected with drugs intravenously during sleep, both REM sleep (purple bars) and dreaming begin earlier than normal. Because in the cat the drug is injected directly into the brainstem, REM sleep begins even more rapidly and lasts much longer than normal. José Calvo and Subimal Dutter have recently shown that in addition to the zone for immediate REM enhancement in the medial pons, there is a zone for delayed—and prolonged—enhancement in the lateral pons. The long-term site is in the vicinity of the cholinergic neurons of the pedunculopontine nucleus (red dots in the bottom right figure on page 138).

Experiments with drugs have also confirmed that the aminergic system suppresses REM sleep. By blocking the noradrenergic system at the same sites that show cholinergic enhancement, scientists can also increase REM sleep! So the push-pull model appears to be correct.

AIM: AN INTEGRATED MODEL OF BRAIN-MIND STATE

Utilizing the data from neurophysiological studies of neurons, it is possible to derive a continuous quantitative estimate of mental state and the state of the brain. Because the two states are inextricably tied together, I will simply refer to them as the brain-mind state. If we can estimate the values of three variables, those values will suffice to tell us the brain-mind state of the individual.

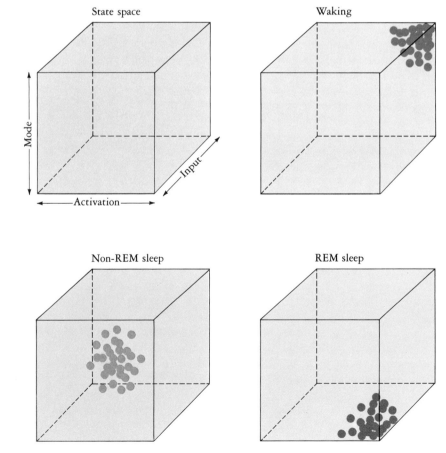

The AIM model may be visualized as a cubic space; each of the three factors determining the state of the brain is represented as a dimension. At any instant the state of the brain can be represented as the position of a point according to the relative intensity of A (activation), I (input source), and M (mode of information processing). During the sleep cycle the cloudlike clusters of points representing the successive, instantaneous values of AIM can be imagined to move from the back right corner (in waking), through the center (in non-REM sleep) to the lower right corner (in REM sleep) of the schematic state space.

Using reticular formation firing rates, we may estimate the strength of activation (A); using sensory and PGO burst cell data, we may estimate the degree to which the input source (I) is external or internal; and using the ratio of aminergic to cholinergic neuronal firing rate, we may estimate the chemical mode (M) of the system, which determines how information is processed. For example, whether or not information is stored in memory could depend on these chemical systems.

The numerical values of the three parameters A, I, and M can be combined in several mathematical ways to estimate the state of the brain-mind. The easiest of these to grasp is the three-dimensional graphic representation shown in the figure on the previous page. Since the parameters vary continuously, the state of the system is estimated by a moving point in the space defined by the parameters.

If all three values are high, as in waking, the points are in the back upper right corner of the state space. As all three values fall in non-REM sleep, the points gravitate toward the center of the space. Then, as activation rises again in REM sleep while input (I) becomes more internal and the mode (M) more cholinergic, the points swing out and down to the lower right corner of the space.

This three-dimensional model not only gives a vivid picture of change in natural state but can be extended to consider abnormal processes, as shown by the figure in the margin. A major advantage of the concept is thus the provision of a common, continuous space in which the behavior of three variables can account for a wide variety of normal and abnormal conditions.

Despite our ignorance about how the sleep process is actually initiated, we have learned a great deal about the mechanisms that mediate non-REM sleep (on the one hand) and REM sleep (on the other) and how the alternation of the two states is programmed by the brainstem. We can, moreover, use the data to account in a preliminary way for the changes in mental state that are associated with the changes in physiology. The result is a quantitative model that combines three factors—activation level, input source, and mode of processing—into a single formula yielding continuous values that estimate the probability of our having wakelike or dreamlike mentation as a function of brain state.

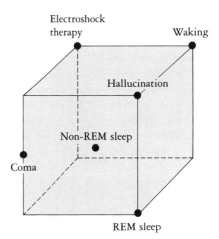

The AIM model may be used to think about the state of the brain in conditions other than sleep. For example, when all three state values fall and the AIM points move toward the left lower front corner, the result is coma. Hallucination would be the effect when the internal stimulus strength increases during waking, and AIM moves to the right upper front corner. AIM moves to the left upper rear corner when external stimulus strength is increased, as in electroshock therapy. Whether or not these assumptions are correct can now be tested experimentally.

Walking along a dirt ⌒

C H A P T E R

7

Dreaming

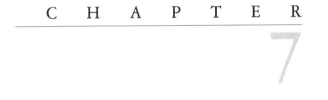
...eet a man traveling on a strange tandem-arranged bicycle.

Our thoughts and dreams during sleep are the subject of sleep psychology. Since neurophysiology has begun to reveal the brain processes associated with these psychological phenomena, scientists are able to study brain processes and mental processes in conjunction. The opportunities provided by our current capability to define and measure both the psychological and physiological dimensions of sleep put the field of sleep research in the center of the age-old issue that philosophers call the mind-body problem.

In this chapter, I will first discuss the mind-body problem and develop a specific strategy for approaching it. I will then examine the evidence that certain characteristics of mind states in sleep correspond to certain characteristics of brain states in sleep, whether the mind state is a dream in REM sleep or unconnected images in non-REM sleep. I will suggest that dreaming, in all its formal aspects, is determined by human physiology: the visual perceptions,

The formal qualities of dreams such as their bizarreness can be analyzed by examining home- and laboratory-based dream reports. This drawing of a double bicycle and its accompanying longhand description come from a dream journal written by a scientist in the summer of 1939.

the strangeness, the illogical thinking, the lack of memory, the emotions, and the lack of insight are all determined by the physical condition of the brain in REM sleep.

I call my own theory of dreaming the activation-synthesis hypothesis in order to focus on two central aspects of dreaming. Activation is an energy concept: in REM sleep, brain circuits underlying consciousness are switched on. Synthesis is an information concept: dream cognition is distinctive because the brain synthesizes a dream plot by combining information from sources entirely internal to itself and because chemical changes radically alter the way information is processed. So the term "synthesis" implies both fabricated (made up) and integrated (fitted together).

SLEEP AS A WINDOW ON THE BRAIN-MIND

Throughout history, thinkers have struggled with the question, How can thoughts, feelings, and perceptions be related to the body? While most philosophers and naturalists have assumed some such relationship, they have not been able to agree on several important questions: What part of the body should they focus on? (Aristotle, for example, favored the heart.) What might the nature of the relationship be? (René Descartes, for example, thought that the mind and the body were two independent entities, which were perfectly synchronized by god and neither of which could effect any change in the other.) And some thinkers were unconvinced that any such mind-body relationship existed at all. (For Plato, the only absolute reality was the idea, the rest being a mere shadow.) Extreme exponents of this position, called idealism, hold that immaterial souls only temporarily inhabit bodies and then enjoy eternal life outside them.

Against the idealistic trends is a tradition called materialism. For the strict materialist, thought is an epiphenomenom of brain activity and is not itself able to act as a cause. Materialism has its own long history: the Ionic Greeks believed that all phenomena, including thought, were caused by the movement of particles; Galen localized consciousness to the brain; and the English philosopher John Locke believed that consciousness resulted from the activity of the brain, which continued without pause even in sleep.

Today, we are in a strong position to construct a general theory of mind-brain interaction. By relying heavily on physiology, we follow the materialist tradition of Locke. By paying equally careful attention to psychology, we are as genuinely idealist as Plato. We may transcend both, however, if we are able to realize that since mind and brain are an indissoluble entity, the study of either aspect is not a substitute for the study of the other. By constantly reminding ourselves of this important point, we may sail between the Scylla of a brainless mind and the Charybdis of a mindless brain.

That is why I will use the term "brain-mind" or "mind-brain" to refer to the functional entity of my interest. In asserting that sleep opens a window wide upon this subject, I mean that, as we pass from waking through non-REM to REM sleep, the states of the mind change so dramatically and reliably in parallel to the states of the brain as to suggest that one or both aspects are the cause of the changes in the other. We may thus be optimistic that we are able to sketch the possible nature of the brain-mind relationship.

While my conviction is clearly more monistic than dualistic, I do *not* think we are yet ready to dispense with a language that discusses mind apart from brain. We still do not understand enough about the brain to understand language itself or to understand the capacity of a physical machine to have an idea of itself. When we have achieved this understanding, we will have evolved a description quite different from any we can give today.

TWO COMPETING HYPOTHESES ABOUT DREAMING

Since the early twentieth century the leading dream theory has been that of psychoanalysis according to Sigmund Freud. In Freud's theory, the key process in dream formation is the psychological disguise of an unconscious wish. During sleep the ego (or self) relaxes its vigilance upon the id (or instinct),

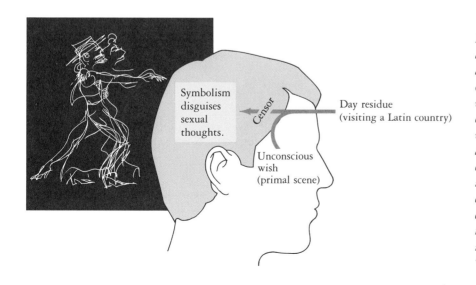

In Freud's psychoanalytic model, my tango dream (see pages 158-159) could be seen as the result of the release from repression of an unconscious wish (such as the visualization of parental intercourse, the so-called primal scene) and its association with the memory of a recent experience (my arrival in Italy, a Latin country, the so-called day residue). In order to prevent the unconscious wish from entering my consciousness (and disrupting my sleep), the censor would disguise and otherwise transform these "latent" dream thoughts into the apparently absurd and meaningless "manifest" content of the tango dream.

allowing forbidden wishes or drives to escape their safe confinement in the unconscious. Were these freed desires to invade consciousness, they would disrupt sleep. A part of the psyche called the censor protects sleep by transforming the true import of the unconscious wishes into symbols whose meaning is hidden from the dreamer (see the diagram on page 145). For Freud, all aspects of the remembered (or manifest) dream content were the result of censorship and disguise of its true content. Thus, the true meaning of dreams was obscure to the dreamer but could be unveiled by the psychoanalyst using an interpretive technique that asked the dreamer to free associate in response to the manifest content. Freud's theory falls into the idealist camp because it ignores the physical condition of the brain entirely.

Neuroscientists now have a greater understanding of the brain than was possible in Freud's day. Believing as we do that mind and brain are inseparable, my colleagues and I at the Laboratory of Neurophysiology at Harvard Medical School decided to take a different approach from Freud's: we decided to look for a way to compare the characteristics of dreams with what we knew about sleep physiology. But which dream characteristics would we focus on?

Given the still very primitive state of our knowledge, we have to be careful not to ask too much of sleep psychology. For example, since we don't know the neurophysiology of language, we can't expect neurolinguistics to help us understand the narrative aspects of dreams. And yet language has been the traditional focus of dream theorists from Artemidorus to Sigmund Freud. Dreams—as stories—are literature. And we have no physiology of literature. Little wonder that studies of dream narratives have failed to achieve scientific status.

But the fact that we can't have it all doesn't mean we can do nothing! Since we now have a picture of brain states in sleep, we might ask ourselves if we have an equally good picture of mind states in sleep. The answer is clearly *no*. Thus, for example, we still lack an adequately detailed inventory of the kinds of illogical thought processes that are present in dreaming. People, including most scientists, have been too occupied with dream *content* to pay attention to its formal aspects.

By the formal aspects of mind states, I mean the nature of the mental processes (for example, is memory good or bad, recent or remote) rather than the content of the processes (for example, what is remembered or forgotten). The formal aspects of a dream include such features as the kinds of emotion felt, type of movement experienced, particular colors seen. They are complemented by the formal aspects of brain activity, which refer to such issues as whether external or internal signals are processed, whether or not the signals are acted upon, and whether or not they are recorded in memory. The content refers to the signals themselves. Evidence suggests that there is a correspondence between the formal aspects of mind states and the formal aspects of brain states. Thus, the study of one may be a source of insight into the other. If we discovered, for example, that attention and memory were weakened in dreaming, we would be better able to ask the brain, "What is it that you are doing less well in REM sleep than in waking?" And if we were to discover that visual image intensity was greater in dreaming than in waking, we might ask the brain, "What is it that you are doing better in REM sleep than in waking?"

Our emphasis upon the formal aspects of dreaming was, in fact, a consequence of our description of the formal aspects of REM sleep neurophysiology through the reciprocal interaction model described in Chapter 6. Once that theory was in place in the mid-1970s, Robert McCarley and I proceeded to develop its psychological counterpart, which we call the activation-synthesis hypothesis of dreaming. Stated simply, the activation-synthesis hypothesis asserts that dreaming is the subjective awareness of brain activation in sleep. The distinctive formal properties of dreams, such as visual perception, uncritical belief, bizarre occurrences, flawed reasoning, intense emotion, and poor memory are all the direct consequences of change in brain physiology. While we know much less about the synthesis of dream content than we do about activation, we can even begin to suggest neurophysiologically realistic explanations of some aspects of dream plot formation.

Some specific assertions of the activation-synthesis hypothesis include the idea that the same thalamocortical brain circuits that underlie consciousness in waking also do so in sleep. In contrast to what happens in waking, however,

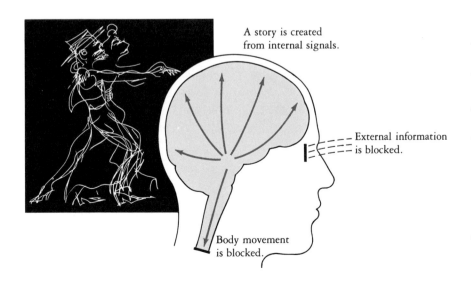

A story is created
from internal signals.

External information
is blocked.

Body movement
is blocked.

According to the new activation-synthesis model, my tango dream is the result of automatic brain activation in REM sleep. The unique and bizarre tango plot is a function of the undisguised knitting together (synthesis) of many disparate but associated memory contents (my love of music, dance, travel, the exotic, and yes, even the sexual). Contributing to the vividness of the dream is the stimulation of my visual brain by signals arising from brainstem motor systems. My sleep is protected from disruption by physiological blockade of both sensory input and motor output.

external sensory input and motor output are both actively blocked during sleep by the neurophysiological mechanisms discussed in Chapter 6. The brain-mind does not process external information; instead, it uses internal signals such as the PGO waves as a source of data from which it synthesizes a dream experience that is as coherent as possible (see the diagram on page 147). The incoherence of dreams is in part related to the chaotic nature of the PGO system and, in part, to the alterations in neurotransmitter ratio that I have described in Chapter 6. The decrease in aminergic neurotransmitters also impairs the memory process, leading to dream amnesia.

HOW DO SCIENTISTS STUDY THE MIND IN SLEEP?

In order to correlate the formal aspects of mental activity and brain activity, we need first to learn what those aspects are by studying the actual mental activity of subjects during sleep. How do scientists study mental states in sleep, whether they be dreams or thoughtlike mental activity?

Unfortunately, scientists never study the mind directly but always through the medium of language. Even when studying the waking mind, they only study the *verbal reports* of consciousness, not consciousness itself. When studying the mind in sleep, they must wake subjects up even to get reports. But then the subjects are obviously only reporting in one state, waking, what they recall of the previous state, sleep. To imagine the difficulties inherent in such reporting, think of your own feelings on being awakened by a phone call in the first two hours of sleep. Disorientation, confusion, and terror are not uncommon. How reliable would you expect reports of antecedent mental activity to be under these conditions?

But let us look on the bright side. Think about lying in bed on a Sunday morning drifting lazily in and out of sleep. On coming out, you remember a dream. Sometimes quite well. Certainly well enough to describe its formal aspects: what sense impressions you had, what movements were made, how you felt, what was strange about it, whether you thought it seemed real, and so on. Mightn't you have as much conviction about such a dream report as about a report on any other aspect of mental life?

I draw this contrast between awakenings that are instrumental (the phone call) and those that are spontaneous (the Sunday morning drifting) to introduce a brief discussion of the difference between dream collection in the sleep lab and dream collection at home. Each method has advantages and disadvantages. In the lab, the physiological reactions of subjects are constantly monitored; experimenters have control because they can wake up subjects at any point in the sleep cycle, and they have the semblance of objectivity. But subjects don't sleep well in labs, and their dreams lack the intimate details and emotional range of home dreams. It is as if the subjects knew they were being watched—which they are! When I myself volunteered as a subject, I became aware of yet another problem with the sleep lab: I was so eager to please the experimenter that it was hard to say, when feeling confused upon being awakened, "I really don't have any idea what was going on." At home, subjects may be more comfortable and hence more uninhibited, but they cannot choose the part of the sleep cycle from which they will awaken, and they have to record the dreams themselves instead of having a tape-recorded interview with the experimenters.

What experimenters would like is the advantages of both and the disadvantages of neither. With the home-based, physiologically controlled awakening made possible by Nightcap, they can now approach that ideal condition. And now they can use both approaches and try to extract the common elements to achieve greater reliability and validity.

My colleagues and I have collected over 1000 sleep lab reports of mental activity. These reports include 73 carefully matched pairs, each from a single

This page is from the scrupulous and lavishly illustrated dream journal kept by a Smithsonian scientist in the summer of 1939. I call him the Engine Man because of his fascination with trains, such as this circus train chugging up a steep, curved grade. His written accounts of this and other dreams have provided data for an analysis of dream bizarreness and other formal qualities of dreams.

subject awakened once in REM and once in non-REM sleep. In addition, we have studied the reports of several individuals who slept at home and recorded what they recalled in dream journals. An illustration from one such journal appears on page 149.

By comparing these reports, my colleagues and I were able to identify some characteristics of mental activity in the different stages of sleep. I give our results in the next few sections. Despite our still limited knowledge of how the brain operates, I attempt to draw some parallels between the activity of the mind and the activity of the brain.

MENTAL ACTIVITY IN WAKING AND SLEEP

One way to understand mental activity in sleep is to compare it with mental activity in waking. While we are awake, we continually exchange information with the outside world. Through our senses we perceive the world, and through our muscles we act upon it. Our sensorimotor behavior is guided by our brain-minds, which maintain a program that orients us and plans our behavior and which also continually analyzes the success of that program. Orientation is crucial to our establishing constant inner representations of time, place, person, and action. Thus, immediately upon awakening, we are aware of who and where we are, what day it is, and what our plan for the day will be. These mental guidance systems are such important and unconscious functions that we take them for granted, but as we shall see, we regularly lose them in sleep.

The ongoing analysis of our behavior encompasses both concrete issues, like whether the weather is appropriate to the clothes we selected, and more abstract ones, like whether the interpersonal climate is favorable to the social goals we mean to pursue. Insight and judgment are critical to these functions. So is logical thought. All are badly impaired in sleep.

In addition to this ongoing background analysis, the waking brain-mind is socking away, in short-term memory, each part of our foreground experience. At the end of any given day, the amount of information we have been exposed to, and the amount we have stored in short-term memory, is astonishing. When James Joyce tried to account for one day in the life and mind of Leopold Bloom, he produced *Ulysses* and was, no doubt, dissatisfied with that monumental document. No one could possibly record, in writing, all the information one brain-mind processes in one day. In contrast, most of us would have little difficulty describing our average daily dream recall on one page, and many pages would be blank.

During waking while we are conducting an analysis of our surroundings, keeping on track with the aid of our orientational compass, and storing bits of information in memory, we may *also*—as if this weren't enough—be imagining things. Some of these imaginations involve review and rehearsal. That is, we call up scenarios of past or future events in order to change these in our favor or anticipate various possible outcomes. Other fantasies are more remote—friends from long ago may pop up—or more abstract—a song, or a design, or an experiment may suddenly occur to us. Some of these imaginings are so fugitive that they go completely unnoticed.

These imaginative mental activities share an important characteristic with our orientational and analytic processing: they become conscious as narrative structures, as words in our head. Just as "Today is Tuesday—a work day—appointments beginning at 8—must hurry" is an example of an orientational narrative, or "In my interview with x, suppose he says y when I suggest z" is an example of a rehearsal. In our fantasies, we may actually visualize encounters with others, down to the details of our behavior and conversation. This sort of mental process is enhanced during REM sleep; while retaining its narrative structure, the imagining becomes even more scenariolike. In the full-blown dream experience, the imagined action has a life of its own; it will run outside of any narrative structure that tries to contain it.

We have only fragmentary details of how human physiology influences these mental processes. As pointed out in Chapter 6, we know that the brain is electrically activated and becomes more so, the more we fix our attention on specific objects or events; we know that our brains perceive the world and command actions through well-controlled sensory and motor channels whose gates are kept open by messages from the reticular formation; and we know that novel stimuli, which invite attention and are more memorable than expected ones, evoke increased firing of the aminergic neuronal systems of the brainstem. While woefully incomplete, this knowledge produces at least a skeleton on which to develop a more muscular theory of the brain basis of consciousness.

MENTAL ACTIVITY IN DROWSINESS AND AT SLEEP ONSET

There are two ways we become aware of a loss of alertness. One is the perception of sleepiness itself: My senses are dulled; I feel tired and have trouble keeping my eyes open; my attention wanders. The other is more subtle: With-

At sleep onset, unformed images may arise as the now unstimulated brain continues to process its own visual data. Drawings of these so-called hypnogogic visions were made by the nineteenth-century French scholar, Hervey de St. Denis. At that time it was reasonably inferred that these "wheels of light" and "tiny little bubbles rising and falling" originated on the retina at the back of the eye. We now know that the formed imagery of dreams arises within the brain itself.

out feeling sleepy, I may notice that I am simply not processing the data I am taking in; my eyes continue to meander over the sentences in my book, but I have no memory of what those sentences said, I am not really reading them.

In this example, my brain waves would slow and increase voltage in the shift toward Stage I sleep. Eye movements decrease. Muscle tone diminishes. Reaction time declines. I might be frankly asleep. Were an EEG recording made of my brain wave patterns, the EEG would show further slowing and perhaps a sleep spindle.

As this process advances, either rapidly (in seconds) or slowly (in minutes), I may have dreamlike mental experiences. Visual images may appear, and I may imagine brief scenarios; convulsive muscle twitches may occur and with them a feeling of falling; I may awaken suddenly, having incorporated a noise or light into a microdream of hallucinatory intensity; on opening my eyes, I may see someone in the room for an instant.

This moment of falling asleep has fascinated students of consciousness for generations. The Romantic poets used their capacity to go in and out of such borderland states to smuggle fantasies from the darkness of their minds' edge into the light of their art. The late-nineteenth-century French physiologist Alfred Maury tried to track the fate of outside stimuli, such as light or noise, across this threshold and evolved theories of their transformation through association into the ideas and images of revery. The surrealists used the freedom from conscious control engendered by these slippings away to discover the unconscious logic of automatic writing. And Marcel Proust detailed his insomniac entries into and stay-in-bed exits from sleep as a part of his extensive study of internal states in the *Remembrance of Things Past.*

What can the modern theorist say about this mental activity at sleep onset? Only that the cortex, being still highly activated but suddenly deprived of external data, may run free, both generating sensory impressions and constructing scenarios out of memory, until the quick ebbing of brain activation makes this no longer possible. This simple model is in a sense a variant of the activation-synthesis hypothesis of dreaming; it enables us to understand why mental activity at sleep onset is dreamlike and yet neither as intense nor as sustained as the dreaming associated with REM sleep.

MENTAL ACTIVITY IN NON-REM SLEEP

Whether or not we have microdreams in descending Stage I, soon either we are utterly oblivious and without any recallable mental content or we enter a most peculiar sleep-thought mode. This latter possibility is particularly likely if our brain has been crammed full of ill-digested information or we do not sleep soundly. Our brain-minds then behave like a cow digesting its meal: they ruminate, chewing the cud of the mental hay but with less than even bovine efficiency. It is as if we can't either get fully off-line or stay productively on-line. We run in place. We whir. We buzz. We may even suffer, so obstinate is our sleep thinking. Many prefer the oblivion provided by sedatives to this cognitive quagmire.

Can the intermediate shades of grey between the white nights of our insomnia and the jet black nights of our oblivious sleep be faithfully tracked by the spectrum of our brain waves? It seems likely. And yet we know of dramatic dissociations between what sleepers report of mental activity and what their EEGs reveal. Some subjects tell us their sleep was fitful and unrestful despite perfectly healthy looking non-REM EEG profiles. And I have awakened subjects from electrophysiologically unequivocal non-REM sleep only to be told "I was not asleep." The EEG may not be a particularly sensitive instrument when it comes to objectifying goodness of sleep and to predicting mental activity in non-REM sleep.

When I awakened one subject from Stage IV, he gave an elaborate, rambling, and garbled report. As I listened to his account of a tank and airplane attack, I looked at his EEG and saw the huge, languorous wave forms of Stage IV sleep! He was sleep-talking. His report was called a dream by judges reading his transcript, but did it occur before or after the "awakening"?

Despite the methodological problems, certain patterns emerge from non-REM sleep lab data. About half the awakenings from Stages II, III, and IV of descending non-REM sleep yield reports of *no* mental content, while about 40 percent yield reports of thoughtlike content. The remaining 10 percent of the reports are of dreams, with formed visual images and fanciful plot structures. But these reports are shorter and less bizarre than the dream reports of REM sleep.

The first two cycles of the night have the most non-REM sleep and the deepest non-REM sleep. Mental activity is, on average, at its lowest ebb. Toward morning, the general activation level of the brain-mind rises. The REM phases become longer, and so do the dream reports. The non-REM phases become shorter and shallower, so that the last two cycles typically contain only Stage II sleep. The difference between mental state features in non-REM and REM sleep should decline just as the differences in EEG patterns did, and, in fact, early morning awakenings from Stage II yield many

more reports of dreamlike mental activity than do awakenings earlier in the night.

Cognitive psychologist John Antrobus proposes that the only significant difference between brain-mind states in sleep and waking is activation level: waking is the most highly activated state, REM sleep less so, and non-REM sleep the least. While I agree that non-REM sleep is the least activated of these states and that this accounts for the shortness of non-REM awakening reports, I have already explained my reasons for postulating two other factors, input source and modulation of neurotransmitter release, to account for the formal differences between waking and the mental states non-REM and REM sleep.

DREAMING AND REM SLEEP

The discovery that REM sleep is the physiological state of which dreaming is the subjective awareness has generated sustained excitement in those scientists seeking understanding of the relationship of mind to brain. Here, at last, was a correlation so strong and dramatic that it promised real progress in establishing a detailed picture of that relationship. With the details of physiology in mind, I now turn to the data on REM sleep from the human sleep lab.

Dream reports from REM sleep awakenings are not only longer than those from non-REM sleep awakenings, but the dreams are more bizarre and more vivid, and they involve more physical activity and emotion. Further-

Drawings such as these from the Engine Man's dream journal can be used to illustrate the system used by my colleagues and me to categorize dream bizarreness. Our system is sensitive to such items as physical impossibility (the dreamer riding above the ground on a flying carpet) and physical improbability (the dreamer's own mechanical forerunner of a modern electronic information retrieval system). More surrealistic features include an automobile bridge that suddenly changes level (discontinuity) and a large truck station in the middle of a flower garden (incongruity).

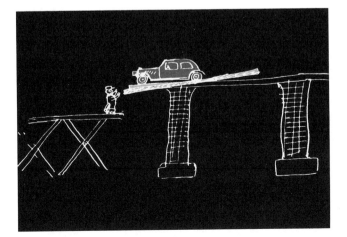

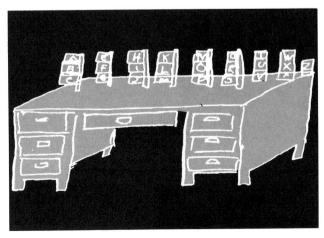

more, these characteristics, as well as recall, are enhanced if the awakening occurs during an eye movement cluster, as opposed to during a lull between eye movement clusters. This means that within each REM period, our dreams become momentarily more intense whenever there occur momentary increases in excitation of the visual, motor, and other systems.

The length of a dream report is another mental state feature that shows a correspondence with sleep physiology. The longer the REM period has been in progress, the longer the retrospective report of dreaming is likely to be. We have all had the experience of awakening with sharp recall of a scene and with full certainty that there were previous scenes that we *can't* recall. Whether or not subjects can remember all the details, their estimates of the time elapsed in the dreaming correlates with the time spent in REM sleep prior to awakening.

The easiest way to account for variations in dream length and vividness is to assume a one-to-one relationship between the formal aspects of a dream and the formal aspects of brain physiology in REM sleep. As we shall see, the evidence for such a relationship is strong.

The first attempt to specify a one-to-one relationship between dreams and physiology focused on eye movements. Howard Roffwarg and his co-workers at the Columbia University School of Medicine recorded subjects' eye movements in REM sleep and then awakened the subjects and tape-recorded their dream reports. Judges scored the goodness of fit between the direction of the recorded eye movements (up-down, right-left) and the direction of the dreamer's gaze inferred from the dream report. A large number of good-to-excellent fits were obtained. Roffwarg's paper on the findings included some rather spectacular anecdotes: a sequence of up-down eye movements was associated with the dreamer's going upstairs, a sequence of right-left movements with watching ping-pong.

These results gave rise to the scanning hypothesis of dreaming, which proposed that in REM sleep there was a one-to-one relationship in real time between eye movement direction and the direction of dream gaze. Simply put, when we dream that our gaze shifts, our eyes move in the direction of the shift. A physiological model of the results would say that the cortex is getting information about each REM and fitting gaze-appropriate images to it *and/or* that the cortex is perceiving the image and then moving the eyes in the appropriate direction. I emphasize the "and/or" because, based on our knowledge of visual processing during the wake state, it is highly likely that both would occur, and they are not, of course, mutually exclusive mechanisms.

Two follow-up studies have failed to replicate the original findings, leading many dream researchers to reject the scanning hypothesis. But because of difficulties posed by the methodology of the study, the odds against finding anything at all are so great that openness on this question is warranted. How, for example, is one to establish exact second-to-second alignment in time between an eye movement recording (which is precise) and a dream report (which is at best only sequentially accurate)? Suppose what feels like 10 seconds in a dream is really 5, or 20, seconds of real time? And, if a single eye movement in a sequence doesn't fit a gaze shift in the subjective report, is the overall correlation to be considered good, bad, or indifferent? This problem may simply exceed the current capabilities of our methodology.

An alternative to the one-to-one model is a global model. Instead of asking if a particular dream scene corresponds to a specific set of eye movements, we can simply look for evidence of a general correlation between the mental state of dreaming and the brain state of REM sleep. Thus, we may look for a correlation between visual intensity or number of gaze shifts, and physiological indices of internal stimulation such as increased frequency of eye movements or of PGO waves. The activation-synthesis hypothesis is an example of a global model.

Apart from its breadth, this global form of the model has other advantages. Since researchers do not look for a second-by-second cross correlation, they do not need to know the exact sequence of eye movements and can dispense with sleep lab recordings. Thus, the model allows subjects to sleep and record their dreams at home. Since dreams recorded at home are more emotional, more sexual, and free of the strong psychological influence of the sleep laboratory, the psychological data can be broadened, deepened, and normalized.

The other advantage is that inclusion of physiological data from animals becomes possible; such data allow us to explore the workings of cells and molecules, a critical step in building more realistic brain-mind models. We need not assume that cats dream, only that the general organization of their brain activity in REM sleep is homologous with that of humans. And since we know of no exceptions to this assumption, we accept it.

WHAT IS DREAMING REALLY LIKE?

When we shift our focus from dreams as stories to dreaming as a mental state, we only postpone for a while considering their content. In what follows, I will explore the formal features of dreams by examining a single home-based report. To demonstrate the utility of the global version of the one-to-one model, I try to give an account of the distinctive psychological aspects in terms of physiology. Then, when I return to the question of content, I may find that psychology will have much less explaining to do than I might otherwise have supposed.

I will base my discussion on a recent dream of mine, which I have called "A Phallic Tango." Here is the transcript from my dream journal entry of October 21, 1987.

"I am at a beach in South America with my wife, Joan, my second son, Christopher, and other unidentified members of my family. It is a beautiful day, and the atmosphere is festive. Suddenly I notice that between us and the water, a group of young people are dancing the salsa and the samba. I hear, with intense clarity, the music of a Latin American orchestra. Then it seems that we are to witness an exhibition of tango dancing. The dance space has become more like a theatrical stage, and a man dressed in a black bolero suit has begun to tango with his partner, a woman dressed in a carmine-colored gown with a tight-fitting bodice and a long, trumpetlike skirt dotted with puffy pom-poms. I admire their slow, sensual, symmetrical moves as they slide, open and closed, toward us. But now, in fact, they are hardly moving at

In my tango dream the partners whirl and dip as is appropriate to that sensual Latin dance step. But dream discontinuities and incongruities soon emerge. First the male dancer's bolero trousers turn into scotch plaid knickers and then—with the dancers suddenly immobile—his elephant trunk phallus picks up the rhythm of the tango. If this is a disguised sexual impulse, it is easy to see through the mask!

all! Yet somehow I attribute movement to their static pose. Then I burst out laughing because I see that the man's trim black trousers have changed into a red-and-white checked pair of Scottish knickers! What is more amusing is that there is no crotch in these pants, so that his prodigious sexual equipment is in full view. And now this part of him—and this part only—commences to twist and writhe to the tango beat. My laughter begins to stimulate amusement in the puzzled and inhibited crowd, who seem not quite to know what to make of this unusual exhibition. Noticing how long and sinuous his dancing phallus has become, I ask someone, "I wonder if he can pick up peanuts with it?" and it suddenly resembles an elephant's trunk. Noticing that a young man to my left is blowing up one of those long, thick, sausage-shaped balloons that we used to sculpt poodles with, I ask for a cerulean blue one and proceed to twist off segments and sail them into the air above the crowd. This clinches their shift in mood from serious dance-watching to circus clown appreciation, and they join me in uproarious laughter. I wake up."

In developing a global model of dreaming, I first ask, "What are the formal features of the dream?" After responding to this question, I will proceed to discuss what those features might imply about both physiology and psychology.

Sensory features

The dream is intensely visual throughout. There is always a scene before my dreaming eyes. The scene is, by the way, in full color; note the *red* dress, the *red*-checked pants, and the *blue* balloon. When needed, although not continuously, there is organized audible sound, including the vivid Latin tango music. I also sense movement throughout the dream. These are all typical dream features or forms. Absent, as usual, are the senses of smell, taste, and pain. Despite the blatantly sexual dream content, there were no erotic sensations.

Motor features

Movement is continuous, although there is a curious shift in its locus. First, I am walking onto the beach; then I stop and notice the people dancing; then the dancing is limited to the tango couple, while I am standing still watching until they, too, stop and I attribute movement to them. With the metamorphosis of the male dancer, the movement returns and is located in his elephant trunk phallus. Up to this point, most of the movement is of lower extremities. Then I twist off and sail the balloons, at which point upper extremities also become involved.

 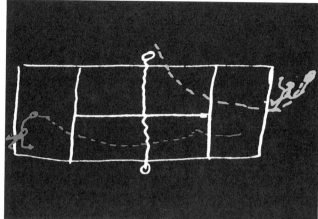

Many drawings in the Engine Man's dream journal represent the scene from overhead in order to define and dramatize unusual movement. The dreamer's car goes round and round in circles in a space impossibly small (left), and both his tennis contestants hit abnormally curved serves into the wrong parts of the court (right).

Cognitive features

Orientation. The dream is typically bizarre: the setting is exotic and remote (I had never been to South America); the characters range from being stable (I am me throughout) through unstable (my wife and son fade) and unidentified (other family members are initially vaguely present) to highly unstable (the tango dancer's metamorphosis). The time frame in this dream is more continuous than is often the case. And so is the action—it is all about dancing. But certain plot features show dramatic discontinuity: the dancer's costume is suddenly, radically altered; the dance itself shifts three times—from young people to the exhibition couple, from there to my mind, and then back again to the dancing phallus. The various kinds of changes are all tied together associatively, but they are nonetheless radical fractures of orientational stability, which would never occur in waking life, even in my fantasy!

Intellectual function. My analytic, critical faculties are in complete abeyance. I accept the illogical transitions without even trumping up bogus explanations, such as "it was clear that the dance couple had just performed in Scotland" to explain the costume change. Even when present in dreams, these ad hoc explanations don't really explain anything.

Insight. Bizarre as my experience is, it never occurs to me that I am dreaming. And yet I am a vivid dreamer, and in my 25 years of professional life, I have recorded hundreds of dreams, all more or less as wild and wacky as this one. So why, world expert that I am, can I not diagnose my own mental state when I am in it? Why have I no insight whatsoever that this is a dream, even though that fact is immediately and convincingly clear when I wake up?

Judgment. In real life, I would not be likely to roar so unabashedly. Nor would I be likely to conspire in such an illogical way to get a crowd to share my perverse enjoyment of life's travesties. After all, why should throwing little balloons up into the air help others understand the humor of the situation if they couldn't already see it? In other words, in the dream my social judgment is distinctly altered.

Emotional features

Humor is the main affect of the dream, and it is vividly expressed in laughter and joking. The dream also contains an element of surprise. There is no anxiety and no fear, however, which is a bit unusual for dreams. Nor are there the negative affects of anger, sadness, or—most conspicuously in this case— shame and guilt. Elation is the second most common dream emotion. Anxiety is the first.

Memory

My recall of this episode is unusually good, probably because the dream occurred at 8 a.m. at the end of a long, deep sleep in a relaxed setting where I had no other obligation than to write this book. Since I was away from home, I was also keeping a journal, which always helps me to remember and to record my dreams. I have no recollection of episodes prior to the beach scene, whose duration I would estimate to be about five minutes.

WHAT DOES A DREAM THEORY NEED TO EXPLAIN?

The foregoing discussion points up some crucial questions that a dream theory needs to answer. Are the dream perceptions a result of sensory stimuli reaching the cortex? If so, what is the source of those stimuli? And what is the source of the movements perceived in the dream? Why is the dream plot subject to such radical discontinuities and incongruities? Why do these so often go unnoticed, and why do we sometimes make transparently fatuous attempts to explain them? Why is our emotional range so frequently shifted to the high side? And why is our memory for dreaming usually nonexistent and yet, suddenly, quite sharp for brief segments? Finally, of course, we will want to ask: what do dreams mean?

I will take up these questions one by one, showing how the activation-synthesis hypothesis can answer them and comparing these answers where appropriate with those offered by psychoanalysis.

The activation-synthesis model specifies how dream images may arise in the brain. Dream vision cannot arise from the outside world because the eyes are closed and visual input from the retina to the brain is also blocked (1). Instead, eye movement command signals (2) arising from the automatic activation of the brainstem saccade generator excite the thalamus and cortex (3), which does its best to fit associated memories to the internally generated stimuli.

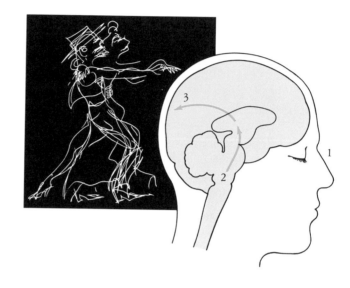

How do the dream images arise?

In REM sleep, the brainstem spontaneously generates signals, such as PGO waves, containing sensory information. In doing so, the brainstem is not responding to external stimuli as during waking but is activating itself. The activation-synthesis hypothesis proposes that these signals are the source of dream images (see the diagram on this page). To the extent that these signals stimulate sensory channels in the cortex in a way similar to stimulation during waking, we naturally elaborate perceptions from them. This sensory mimicry explains the peculiar overrepresentation in my tango dream of certain modalities—especially vision and audition—and the underrepresentation of others—like taste, smell, and pain. Thus, I hallucinated the beach, the dancers, and the balloons because my visual system was kept activated and stimulated. I did not experience taste, smell, and pain apparently because these systems were not activated. Psychoanalysis would ascribe these dream sensations to a regression away from the direct expression of an unacceptable unconscious wish that is the driving force of the dream.

Why is dream movement continuous?

Another brain system that is activated in REM sleep is the central motor pattern generator in the brainstem. When in waking we decide to walk, we voluntarily trigger this system, and it regularizes our gait so that we no longer have to think about it until we want to stop. And when this part of the brain is electrically stimulated in animals, they walk automatically. I suggest that the activation of this system in REM sleep explains how we often perceive ourselves to move in our dreams as if unwilled. The movement is fictive or hallucinatory, of course. No movement actually occurs because the motor output is blocked at the level of the spinal cord, as discussed in Chapter 6. Not all fictive dream movement need be considered automatic and involuntary, as other parts of the motor system are also turned on, for example, the cortex. In my tango dream, I imputed movement to the temporarily immobile dancers by an act of will. And, later, I threw the balloons. A more detailed view of the events in the brain is shown in the diagram on this page.

Psychoanalysis attributes the body's immobility during dreaming sleep to motor paralysis without concerning itself with the cause. As with sensory sensations, it treats the sensation of movement as derivative of the process that protects sleep by disguising the unconscious wishes.

Why is dream cognition impaired?

According to the activation-synthesis hypothesis, the abundance of bizarre activity and the absence of insight and judgment are positive and negative

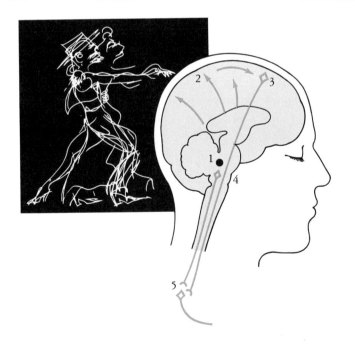

The activation-synthesis model suggests how motion perception may be simulated in dreams. Midbrain motor pattern generators (for walking, running, or dancing the tango) are activated (1), and the command signals are relayed to the cortex (2), where movements in keeping with the dream plot are hallucinated. For example, I throw poodle balloons into the air after imagining a dancing elephant trunk phallus. All voluntary (3) and automatic (4) commands to move are blocked by inhibition (5) of the spinal cord neurons that send motor commands to the muscles. So I don't really throw poodle balloons, although the illusion of doing so is convincing.

aspects of dream thinking that complement and augment each other. Thus, the imaginative and fanciful synthesis of disparate elements proceeds because it is unchecked by the logical rules of waking, while logic fails because it cannot cope with the proliferating images. In terms of brain physiology, something must be added or something subtracted to explain these changes in cognition.

The most obvious and unequivocal loss is of information from the world outside our own heads. Stimuli from that external world are incorporated into dreams only with difficulty and play essentially no effective role in triggering or structuring dreams. The REM sleep brain-mind is thus deprived of the continuous, orderly stream of external information that contributes to the sense of continuity and context and contiguity that orients us in waking. But that subtraction is not enough to explain the disorientation we experience in our dreams, because, when placed in sensory isolation, humans maintain their orientation for many hours simply by thinking.

The brain must have lost its compass as well as its bearings. What else is missing? The most conspicuous neurophysiological debit of REM sleep is caused by the cessation of firing by the aminergic neurons of the brainstem, which could well cut back the supply of aminergic neurotransmitter throughout the brain. Studies of associative learning in animals with simple nervous systems have shown that aminergic neurotransmitters (serotonin in the sea snail and dopamine in the chick) are necessary for such learning to take place. The activation-synthesis hypothesis speculates that in REM sleep the loss of

A striking feature of most dreams is loss of insight. (I can't recognize the tango visions as illusory, although as soon as I wake up this is no problem.) A related deficit is my loss of self-reflective awareness, that splitting of consciousness by which I observe my own behavior—and even my own thoughts—when awake. The activation-synthesis hypothesis ascribes both deficits to two factors: the disorienting effect of the blockade of external ("reality") data and the shutdown of modulatory influences on the forebrain from the aminergic neurons of the brainstem.

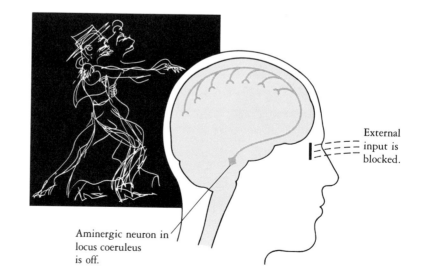

External input is blocked.

Aminergic neuron in locus coeruleus is off.

neurotransmitter chemically alters the brain-mind so that it processes information in a different way. And not only are our thought processes different, but we lose our sense of self-awareness and our critical perspective on these radically altered thought processes.

The sensory signals generated by the brainstem, while similar to wake state stimulation, have significant differences as well: many sensory channels are stimulated simultaneously and others not at all; movement is activated automatically and then not at all; both the sensory and motor systems undergo unpredictable flurries of excitation.

So the tango dancers speed and slow. New sensory data suddenly provide the necessary substrate for the music, for the checked pants, for the balloons. The sensory structure and the pace of the dream are thus determined and changed by these rhythmic stimuli perhaps arising from the brainstem, and the stimuli are fit into the plot as best they can be. If my taste brain was suitably stimulated, I would hallucinate, perhaps, enchiladas!

The dream is held together by a remarkably strong skein of thematic threads. The dance is the narrative frame, but associative freedom is great within this constraint, a tribute to the brain-mind's combinatorial talent.

To explain such dream bizarreness, psychoanalysis assumes that the nonsense is only apparent and that it was engendered in an effort to disguise the dream's true meaning. But how and why the dreaming brain-mind accomplishes the transformation and accepts it is not spelled out. The activation-synthesis hypothesis sees bizarreness as the normal and undistorted consequence of altered information processing; in contrast to psychoanalysis, it does not postulate a disguise mechanism.

Why is the emotional spectrum shifted in dreams?

We know too little about how specific emotions are mediated by the brain to explain why certain emotions, and not others, are characteristically seen in dreams. It may simply be that the disinhibited and hyperexcited state of the brain, in combination with the chemical alteration, supports a shift to more intense emotions. It may even be that the brain-mind, made uncritical by the processes described above, is reacting to its own perverse creations with a mixture of amazement and delight. Although fear, anxiety, and the sometimes extreme aggression of dreams are more difficult to explain, we might view them as part of a response to the dream plot. Thus, in some dream plots I may find myself pursued—and so reasonably frightened—or obliged to fight and therefore to suffer or to inflict harm on my pursuer. It may also be true that such emotions are responses to automatic activation of subcortical rage and fear centers and that the stories are designed to explain the emotions. In any case, *these* emotions are the ones that are there. Although we don't know why, sadness, guilt, and shame are considerably less common.

Psychoanalysis would say that the *real* emotions are the depressive ones, which aren't experienced because the dream is defending the conscious mind against them. That explanation might work for elation and humor (versus depression and shame), but it certainly won't work for anxiety and aggression. In fact, psychoanalytic theory has always had difficulty with the obvious unpleasantness of so many dreams. The activation-synthesis hypothesis has no such problem, since dream feelings, like dream plots, are seen as the undisturbed product of the autoactivated brain.

Why is memory for dreams defective?

If we do not awaken from a dream, recall is likely to be nil. Almost all our dreams are unremembered—the figure is at least 95 percent and probably closer to 99 percent. Instead of attributing dream forgetting to active repression, as psychoanalysis does, the activation-synthesis hypothesis posits a simple amnesia caused by the subtraction from the brain's chemical repertoire of the molecules needed to convert our immediate or short-term memories into longer ones. The molecules would be the aminergic neurotransmitters that are cut off in REM sleep. When we awaken, the noradrenergic and serotonergic neurons turn on and give our brains a shot of these transmitters. If a dream experience is still encoded in activated networks of neurons, it can be reported, recorded, and remembered.

INTERPRETATION OF DREAM CONTENT

According to the activation-synthesis hypothesis, dreaming is the straightforward and unaltered subjective awareness that results from automatic brain activation in sleep. What are the implications of this view for dream interpretation? For all their nonsense, dreams have a clear import and a deeply personal one. Their meaning would stem, I assert, from the necessity in REM sleep for the brain-mind to act upon its own information and according to its own lights. Thus, I would like to retain the emphasis of psychoanalysis upon the power of dreams to reveal deep aspects of ourselves, but without recourse to the concept of disguise and censorship or to the now famous Freudian symbols. My tendency, then, is to ascribe the nonsense to brain-mind dysfunction and the sense to its compensatory effort to create order out of chaos. That order is a function of our own personal view of the world, our current preoccupations, our remote memories, our feelings, and our beliefs. That's all.

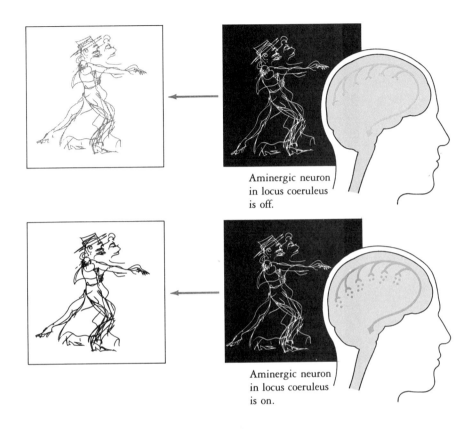

Aminergic neuron
in locus coeruleus
is off.

Aminergic neuron
in locus coeruleus
is on.

Top: *Vivid dream imagery quickly fades if there is no awakening during REM because the aminergic neurotransmitters necessary for learning and memory are not available to help record the experience.* Bottom: *Arousal during REM sleep activates the aminergic system, facilitating recall of the antecedent dream imagery.*

As for deciding exactly why one dream plot is chosen rather than another, or why the plot turns in this direction rather than that, I dare say that no one will ever have an entirely satisfactory answer, for that would entail the capability of predicting a specific dream of a specific subject in a specific REM period. The activation-synthesis hypothesis implies a view rather like that taken by artificial intelligence theorist Douglas Hofstadter, according to whom the number of possible dream plots are infinite and, hence, by definition unpredictable. The reader wishing a more detailed discussion of this topic should consult my recent book, *The Dreaming Brain*.

PRACTICAL CONSEQUENCES OF THE NEW PSYCHOLOGY OF DREAMING

Even if the new findings make our approach to interpreting dreams more modest, can dream psychology still be useful in helping people to function better? And if dreaming can be understood in terms of normal physiology, can we perhaps also apply the emerging model to the understanding of mental illness? I think we can do both.

In dreams, the normal brain-mind quite regularly mimics the most flagrant manifestations of mental illness. Thus, the sensory perceptions are akin to hallucinations; the false beliefs akin to the delusions of psychosis; and the aberrant thinking, memory loss, and confabulatory quality akin to the delirium and dementia of organic brain disease. An adequate account of the mind-brain in dreaming could be a preliminary but possibly informative model of how mental illness works.

That this nightly madness is normal makes us wonder. Could mental illness occur without any structural defect of the brain, for purely functional reasons? Such as loss of a loved one? Such as ambiguous or incompatible demands from important people? If true, this insight would, in turn, broaden our view of functional mental illness (such as the anxiety states) to include a dynamic organic basis. In REM sleep, small and temporary shifts in the balance of neuronal chemistry result in major changes in mental functioning; the same could be true of anxiety states. These states are not "all in your mind"; they are also states of the brain.

The new psychology of dreaming could help all of us, mentally ill or sane, function better. I think that dream psychology can help us to demystify mental life, including dreaming, by acknowledging that it is normally bizarre. This recognition promotes sympathy for those who suffer in waking from the dreamlike states of mind most of us confine to sleep. Having thus enlarged our scope of the normal, we also shed the peculiar modern, psychoanalytic tendency to view even the normal as somehow neurotic. (For Freud, dreaming was like a neurosis in that all mental content of both states was a symptom of pathologically repressed wishes.) Of course, all of life, including mental life, is a set of compromises, but to apply a disease model to this compromise is as peculiar as applying it to dreams. We would do better to discuss alterations in functional states of the brain-mind in a more neutral way.

If the conflicts in dreams are transparent and not disguised, then dreams can be read directly, without decoding and without an interpreter. This could save considerable time and money especially if the interpretations we would have obtained from a psychoanalyst are based on a flawed theory.

We now know from modern experimental work that dreams are easily recovered. The sleep lab, the Nightcap, even an awakening accomplice or alarm clock make it possible to obtain dream reports in one or very few nights. We also know that the consciousness of dreaming can be increased by training in autosuggestion. Subjects have even learned to signal from REM sleep their awareness of dreaming by using an eye movement code.

Since we can learn to be conscious of our dreams while they occur, we can also learn to influence them, that is, shape or change the plots at will. Some modern psychotherapists enable nightmare-ridden patients not only to control their fear of dreamed assailants and so to sleep more comfortably, but also to raise self-confidence and esteem dramatically by actively mastering dreamed threats. Dreaming is a malleable mind-brain state, whose very plasticity can be the means by which individuals can change their fundamental hypotheses about self and world. This view is close to that of the early nineteenth-century Romantics, such as Coleridge and Blake, who saw altered states of mind as liberating and as vehicles to creation. When demystified and depathologized, dreaming is surely an avenue for both entertainment and creativity.

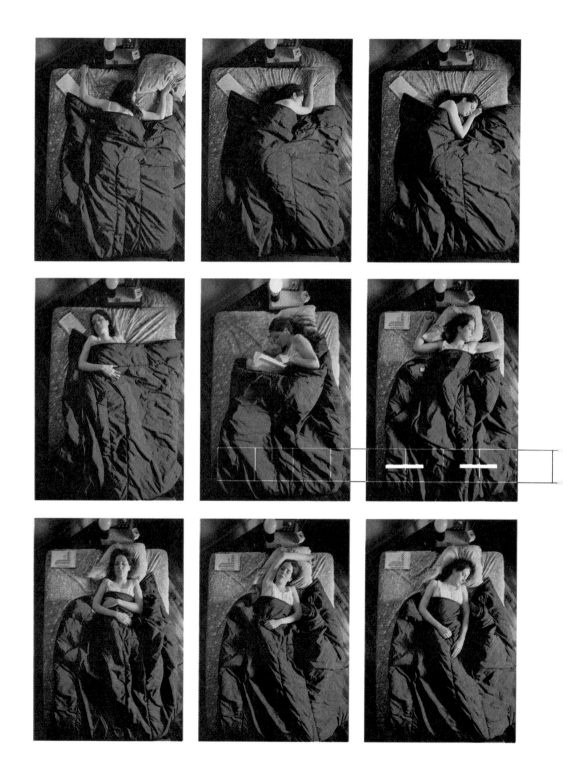

8

Disordered Sleep

Among the directly observable signs that the brain-mind is not quiescent in sleep are those phenomena that have caused people discomfort during sleep since the beginning of recorded history—phenomena such as frightening dreams and nightmares, sleep talking, and sleepwalking. In this chapter, I will show how the reciprocal interaction model of sleep cycle control can be used to help us understand these and other nocturnal phantoms. I will first sketch how the normal dynamics of the sleep cycle could become exaggerated or unbalanced so as to produce too much or too little sleep or the wrong kind of behaviors in sleep. I will then consider a few specific sleep disorders in more detail. Since the reciprocal interaction model is unproven and we do not yet fully understand the neurophysiology of sleep, please look upon the discussion of mechanisms that follows as hypothetical. But the descriptions of the disorders themselves are fact.

Time-lapse photography and video studies objectively document the tossings and turnings of poor sleep. A numerical measurement of the "restfulness" of sleep can be obtained by dividing the number of frames without movement by the total number of frames. For the movement profile above, which records a different sleeper from the one in the photo, the measurement is 12 divided by 29, or 0.4.

THE NEURONAL SEESAW MODEL AND ABNORMAL SLEEP

This aminergic-cholinergic seesaw model of normal and disordered sleep behaves differently if either the weights or the position of the balance point are changed. In normal waking, the weights that represent the instantaneous output of the aminergic (A) and cholinergic (C) systems are arbitrarily taken to be equal, and the balance point, representing the sensitivity of the brain to these outputs, is centered. The normal sleep cycle is an expression of simply altering the outputs of the neurotransmitter systems. Insomnia and hypersomnia can occur if either the weights are changed, as in acute or situational anxiety, or the balance point is altered, as in chronic or constitutional depression.

The reciprocal interaction model presented in Chapter 6 has as its central concept the opposition and continuously shifting balance of two classes of neurons: one excitatory and cholinergic, the other inhibitory and aminergic. If this model is correct, then during normal sleep the levels of activity in these two populations of neurons shift back and forth like a seesaw.

The regular up-and-down motion of a seesaw depends on the weight of the individuals at each end and on the placement of the moving board upon its fixed base. If one individual is heavier than the other, the seesaw is imbalanced and may become stuck unless the position of the board is shifted in a compensatory way. In this analogy the weights of the users represent the changing levels of aminergic and cholinergic neurotransmitter, and the placement of the board represents the mean level around which the levels of the neurotransmitters fluctuate. Thus, if there is too much aminergic force or too much cholinergic force, the balance between those two forces will be shifted: although the relative amounts of force will continue to shift back and forth, the mean level will shift toward one side or the other. As a result, the brainstem oscillator

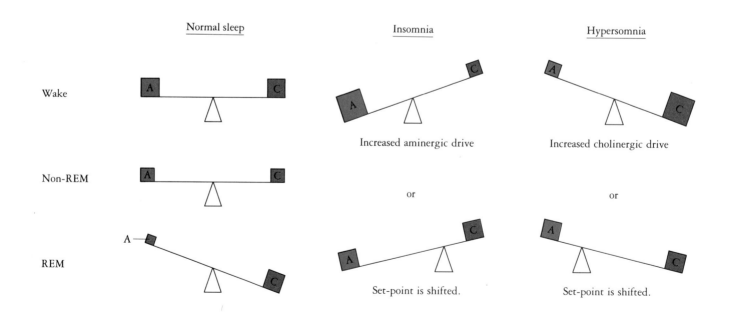

will be more likely to get stuck in the wake or sleep position. The set-point determines when the system will switch from one state or another. Oversimplifying, I consider the set-point to be the ratio of aminergic to cholinergic neuronal firing level (factor M in the AIM model of Chapter 6). In reality, the set-point is determined by a much more complex set of factors. The diagram on the facing page shows how changes in set-point and neurotransmitter force can change our sleep.

The successful use of a seesaw also requires proper timing of the upward spring that shifts the balance in the opposite direction. In the brain, timing errors are also possible: any one of the multiple subsystems under the control of the neuronal seesaw can be sprung loose too early or held down too long. For example, walking and talking are perfectly normal waking behaviors, but we expect them to remain suppressed in sleep. When they pop up during sleep, it is because the relevant motor systems of the upper brain have become excited before the quelling action of inhibition can be instituted in the spinal cord.

In summary, the sleep cycle control system is liable to two major kinds of dysfunction. Set-point errors can lead to too much or too little sleep. Problems with timing can cause normal behaviors to occur in the wrong state.

TOO MUCH SLEEP

You learned in Chapter 5 that when it comes to quantity or quality of sleep, variability across and even within individuals is the rule. Thus, some people are constitutionally long sleepers who need as much as 10 or more hours of sleep to feel rested. Sometimes, however, the tendency to fall asleep suddenly and markedly increases, interfering with the individual's routine. Such cases of sudden hypersomnia, or excessive sleep, may signal the presence of a sleep disorder such as narcolepsy or an emotional disorder such as acute depression.

For both the constitutional and acquired hypersomnias, one possible explanation is that the ratio of aminergic to cholinergic neuronal activity is decreased, with the set-point being lowered (see the diagram on the next page). Since the aminergic side of the oscillator is normally high during the wake state, any process that lowers the level of activity in this system can lead to an increase in the tendency to fall asleep. Such a lowering process perhaps occurs when we are exhausted or during periods of chronic stress. In constitutionally long sleepers, in narcoleptics, and in some cases of depression, genetic factors may have set a low level of aminergic activity. Increased need for sleep also would result from an excessive increase in the level of activity of the cholinergic side of the oscillator.

TOO LITTLE SLEEP

Our subjective experience suggests that insomnia is fueled by an inability to turn off those parts of our brain that are concerned with thinking or feeling. If we are anxious, wound up, or preoccupied, our brains simply cannot enter into sleep easily. Insomnia is the most commonly experienced sleep disorder. Its prevalence may mean that the design of the reciprocal interaction system is intrinsically problematic.

The sympathetic nervous system mediates arousal and anxiety through the sustained activity of aminergic neurons. That being the case, it is possible that the physiological basis for insomnia is *over*activity of the aminergic system, *under*activity of the cholinergic system, or both (see the diagram on this

Top: *To show how acute change in the output of the aminergic system could lead to insomnia, I represent the strength of that system as the height of the red line. When it is raised—as by anxiety—strength of the cholinergic system (violet) decreases. As a result, the change in balance that is critical for sleep to begin takes longer. Even when sleep does occur, it is shallow, and small shifts in strength are enough to tip the balance and cause frequent awakenings.* Bottom: *If the aminergic system declines in strength (or the cholinergic system increases in strength), the difference in strength between the two neurotransmitters is reduced. The sleeper then shifts to a deeper, more prolonged state of sleep (or hypersomnia).*

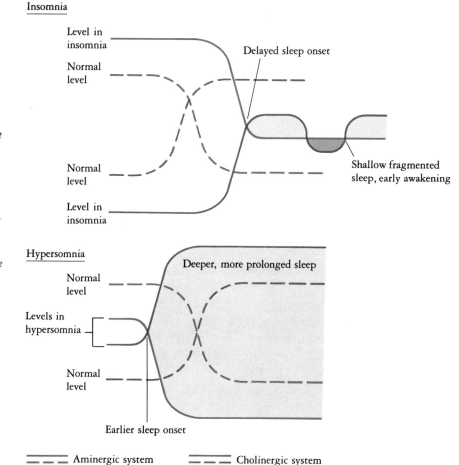

page). As we lie abed sleepless, the neuronal underpinnings of our continuously buzzing thoughts keep the cortical activity levels of our brainstem aminergic neurons too high, and the high activity level in turn keeps our thoughts anxiously buzzing. Sleep becomes increasingly unlikely until fatigue or relaxation breaks the vicious cycle. Later in this chapter, I will discuss ways of overcoming insomnia.

We know just how delicate the balance of the reciprocal interactive system is, since even mild stress or trivial preoccupations can upset it at sleep onset. Moreover, if the imbalance persists for more than even a few minutes, it often gets worse. Thus, many normal sleepers report that if they don't fall asleep within ten minutes, their sleep onset is likely to be delayed for up to an hour. It is almost as if they had to wait for a new brain cycle to occur in order to seesaw down with the now-descending aminergic level into the trough of non-REM sleep.

INAPPROPRIATE BEHAVIOR IN SLEEP

The unity of our personality and behavior depend on the coordination by the brain of myriads of discrete neuronal systems. This unity is disrupted in a group of sleep disorders called parasomnias, whose victims display inappropriate and unwanted behavior during sleep. I now examine how the parasomnias could result from one neuronal system proceeding more rapidly than another into a subsequent state. Timing difficulties may occur in conjunction with sleep onset, transitions from non-REM to REM sleep, and awakenings from non-REM or REM sleep.

If, for example, the upper brain systems underlying thought and feeling successfully enter the sleep state but our motor system remains excited, we might experience at sleep onset a dramatic muscle twitch. The jerk of our limb or trunk may arouse us from sleep or be incorporated into a microdream as a sensation of falling.

Because our cortical motor systems are activated during REM sleep, at each transition from non-REM to REM our brain must perfectly synchronize the activation of upper motor neurons, which command movement, with the inhibition of the lower motor neurons, which execute movement. Failure to do so may result in sleepwalking, sleep talking, and tooth grinding.

On arousal from REM sleep, the forebrain neuronal systems underlying perception are sometimes reactivated before the spinal motor apparatus can reinstate muscle tone. The result is sleep paralysis, an inability to move although conscious and aware of surroundings, which has been experienced by many normal individuals. Another possibility is that the individual on arousal from REM sleep gains a partial awareness of the outside world while the internal perceptual activity of dreaming continues. If so, the individual may perceive a nonexistent figure in the bedroom.

Confusion is often experienced on awakening, and following sudden arousal from Stage IV sleep, we may be particularly disoriented. So detached is our cerebral cortex from the influence of the aminergic neuronal control on which its normal function depends that the brain may take a full five or ten minutes to restore even partially effective cognitive and perceptual functions. In addition, Stage IV sleep occurs near the low point of the circadian rhythm of body temperature, so the brain is at a low ebb of energy as well.

NARCOLEPSY

One of the most common and interesting hypersomnias is narcolepsy, a disorder characterized by irresistible sleepiness. Those afflicted may feel sleepy for the entire waking period, or they may experience dramatic and sudden sleep attacks lasting a few seconds to a minute or two. Some unfortunates experience both forms of the disorder. During a sleep attack, narcoleptics may lose muscle tone, a condition called cataplexy. As a result, they either experience weakness or are unable to maintain an upright posture. Both the sleep attacks and the cataplexy may be precipitated by strong emotion, especially surprise or laughter. There is nothing more ironic and poignant than to hear a narcoleptic patient describe the humorous situations that provoke his attacks and then see his amusement at his stories generate a new attack.

The cataplexy is caused by an atonia that is akin to that of normal REM sleep. Indeed, in these attacks, the REMs can often be directly observed, and the recordings on this page illustrate the coincidence of cataplexy and REMs. These facts suggest that narcolepsy is an abnormality of REM sleep control; that is, the REM sleep generator is not being suppressed during wake periods.

The sudden onset of REM sleep can be recorded in many patients with narcolepsy. Muscle tone (EMG) is suddenly abolished, while eye movements (EOG) are simultaneously generated.

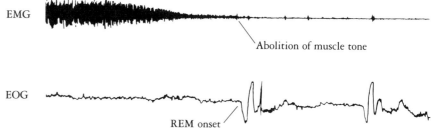

EMG

Abolition of muscle tone

EOG

REM onset

The lack of suppression could stem from a weakness of inhibitory control exercised by the circadian clock in the hypothalamus or from an increased strength of the cholinergic side of the non-REM/REM sleep oscillator or both.

The increased propensity for REM sleep to occur also expresses itself in nocturnal sleep. In narcoleptics, REM periods may occur at sleep onset and may last as long as 20 minutes. During this time, the sleeper may have intense dream experiences. Since the REM sleep process is building during the wake state—causing the excessive daytime sleepiness—it is little wonder that a dream image may be generated and experienced as an hallucination before the narcoleptic loses awareness of his surroundings. Narcoleptics also have difficulty with sleep-wake boundary blurring on waking from REM sleep. In particular, they may experience that difficulty in regaining control of their muscles that I have called sleep paralysis. Sleep onset hallucinations and sleep paralysis may result from a dissociation of REM sleep components, such that the control of the cortex (and conscious state) is uncoupled from the control of muscle tone.

Clinical research has revealed that the sleep attacks of narcoleptics occur periodically at 90- to 100-minute intervals during waking. Thus, their frequency coincides with that of REM periods in nocturnal sleep. This supports the idea that normally during waking the non-REM/REM sleep oscillator is simply held under inhibitory restraint. The periodicity of the attacks is in addition useful to narcoleptics, in that it gives them a sort of "temporal immunity": following a full-blown sleep attack, the individual is likely to be attack free for at least an hour. Indeed, some of my narcoleptic patients have found that, by taking a nap at the time of an expected attack, they can both avoid the attack and have a period of sustained alertness on waking. This is particularly important for individuals who must operate motor vehicles or perform work in dangerous situations. For example, one of my patients, an apple grower in New England, found that following a nap he could safely work for 60 minutes on the high ladders he uses at harvest time.

Narcoleptic symptoms can to some extent be prevented by medications that have the effect of changing the balance of the non-REM/REM sleep oscillator. Traditionally, this has meant using amphetamine stimulants such as ritalin. More recently, tricyclic antidepressants, which are less addictive, are also being used.

Amphetamines control narcolepsy by increasing the drive of the aminergic system, which inhibits REM sleep. Tricyclic antidepressants probably work by acting on both aminergic and cholinergic neurotransmitter systems. First, the drugs block cells from reabsorbing already released norepinephrine; thus, more norepinephrine is present in the synapse to restrain the REM-on side of the brainstem oscillator. Second, despite their name, these antidepressants possess powerful anticholinergic effects, which can directly suppress the cholinergic REM-on neurons of the oscillator.

DEPRESSIVE HYPERSOMNIA

Chapter 4 discussed findings suggesting that long sleepers may be somewhat lethargic and even moderately depressed. Now I note that whereas many clinically depressed patients usually have trouble sleeping, others feel tired and want to sleep a lot. Yet early morning awakening is one of the main signs of depression, indicating that in depression the sleep drive is shallow and weak.

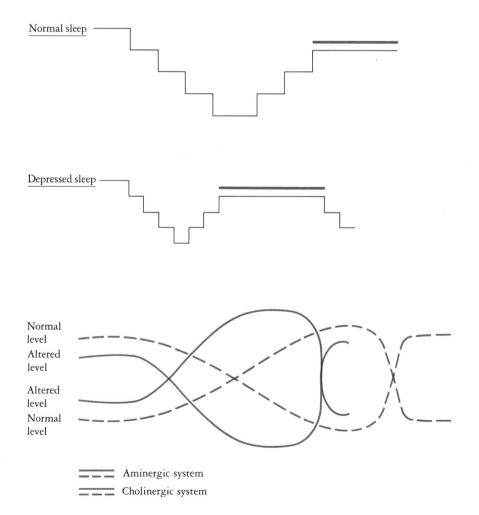

Owing to the same changes in aminergic-cholinergic balance already well-described for hypersomnia, many depressed patients experience an earlier onset of REM sleep during the first sleep period. The first REM period may also last longer and be more intense than normal. This process is reversed by antidepressant medications that strengthen the aminergic and weaken the cholinergic systems.

It is as if the amplitude of the rest-activity rhythm were flattened by the depressive process so that extremes of both rest and alertness are lost. There is also evidence that in depressive hypersomnia the circadian rhythm is phase advanced, which would help explain why some depressed individuals feel tired in the early evening and wakeful in the early morning. For this reason, enforced shifts of rhythm have been found useful in treating depression. The paradox of finding both hypersomnia and insomnia in some depressed individuals may be resolved in the future when we better understand the interaction between the circadian and the non-REM/REM sleep oscillators.

The hypersomnia of depression may well be the result of an increased REM sleep drive. In many depressive patients, REM latency is typically shortened and REM periods are longer and more intense. Significantly, REM sleep deprivation reduces depression in some patients, and naps containing REM sleep may acutely worsen depression. These symmetrical observations strongly suggest a deep causal link between sleep and mood.

Depressive hypersomnia has been ascribed to hypersensitivity to cholinergic neurotransmitters and, alternatively, to decreased sensitivity to aminergic neurotransmitters. As shown in the diagram on the facing page, the reciprocal interaction model suggests that the hypersomnia and the depression are the result of diminished aminergic and/or increased cholinergic force in the non-REM/REM sleep oscillator. Since sleep disorder has always been known to be depression's bedfellow, it is not surprising to learn that they may have a common origin.

THE INSOMNIAS

Individuals suffering from the insomnias have difficulty initiating or maintaining sleep. After a bad night, they feel fatigue and cannot concentrate well. But because sleep length is so variable and sleep problems of this sort are so common, it becomes difficult to say at what point too little sleep should be considered abnormal. Most of us experience difficulty in falling and staying asleep when we are excited, or worried, or missing a loved one. If prolonged or intense, these stress-induced insomnias may become the excessive excitement of mania (in manic depressive psychosis) or depression (the down side of the same disorder). Conversely, individuals predisposed to these mood disorders, or to panic anxiety, or even to schizophrenia, may find their sleep disturbed by their clinical symptoms only to find their clinical symptoms are made even worse by their poor sleep. The common thread in this continuum is anxiety, an exaggerated and unpleasant form of arousal that arises from within (in some people) and from without (in most of us) at one time or another.

We can deliberately induce the relaxation response by controlling mental activity and muscle tone. By consciously relaxing the muscles of the trunk, neck, and limbs, we can reduce the excitatory drive from the muscles upon the aminergic and other arousal neurons of the brainstem. The sleep-promoting shift is further enhanced if we consciously replace preoccupying and emotion-laden thoughts with peaceful visions and the neutral content of a mantra or a sheep count. Our peaceful thoughts will reduce the excitatory drive from cortical neurons. Because our neurophysiology is under conscious control, the relaxation response is an example of the mind affecting the body.

Sleep neurophysiology not only helps us understand what is going on physically inside the brain as we toss and turn but also how to reduce insomnia's unwanted effects. Imagine that when we worry the neurons of our cortex and the rest of the forebrain are activated. These activated neurons in turn excite the reticular and aminergic neurons of the brain stem. By tipping the neuronal seesaw toward the aminergic side, our worries set off an internally sustained drive toward waking.

Since an integral part of arousal is readiness for action, our motor systems are also activated when we feel anxious. We become tense and even visibly fidgety. This reaction means that the brainstem arousal systems are also driven by excitatory signals from the musculature, which add their effects to those of the cortex in keeping aminergic and reticular neuronal levels active and above the threshold for sleep.

This twin process can be reduced or even reversed by the relaxation response, a simple technique drawing upon ancient Buddhist traditions and

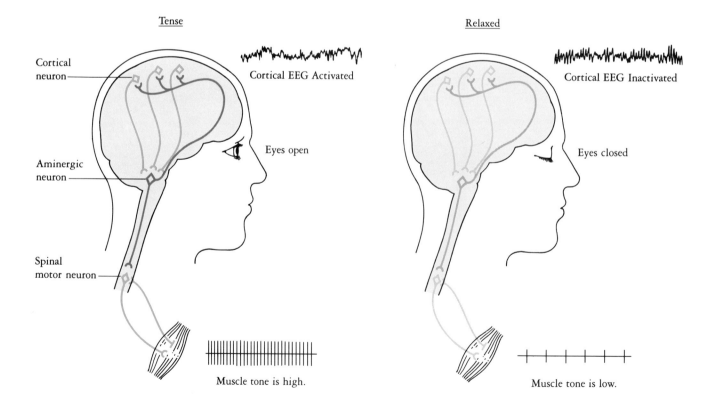

Tense — Cortical neuron — Aminergic neuron — Spinal motor neuron — Cortical EEG Activated — Eyes open — Muscle tone is high.

Relaxed — Cortical EEG Inactivated — Eyes closed — Muscle tone is low.

documented by modern physiology. By an act of conscious will, we may substitute neutral, even nonmeaningful thoughts (like the word *one*, or *omm*) for our preoccupations. To aid this process, we may visualize a pleasant scene (like the mountains or the ocean). As we meditate, the activation patterns of our brain shift both in locus and in intensity, giving our brainstems a chance to quiet down.

The process may also be helped by progressive muscle relaxation. First we relax the toes, then the feet, next the calves and thighs, and so up the trunk, until the upper limbs also relax and even our breathing slows. By these means we are guiding our brains and bodies in the direction of sleep. The illustration on the facing page shows how the relaxation response sets off changes in the brain that encourage sleep.

Although not a panacea, this approach is invariably of some use as part of a larger program including exercise, counselling, or psychotherapy. Sometimes medication is necessary, but even then, these simple active techniques should not be forgotten.

ALCOHOLIC INSOMNIA

Alcohol has profound short- and long-term effects on the quality and quantity of sleep. During the first part of the night following alcohol intake, the REM phase is suppressed. Sleep may be unusually profound, or even semicomatose if the dose of alcohol is high enough. Use of alcohol suppresses posture shifts that are normally triggered by sleep phase transitions; the resulting compression of the radial nerve in the upper arm may contribute to the "Saturday night paralysis" that can lead to permanent hand weakness in alcoholics. Insomnia does not occur until later in the night, when the aldehyde by-products of alcohol breakdown accumulate. Sleep becomes fitful, and the interruptions of sleep result in additional REM deprivation. A hangover typically includes feelings of sleepiness and fatigue, which alcohol may temporarily reverse only to further intensify.

In chronic alcoholics, the long-term REM sleep deprivation may be a significant factor in the ultimate development of the alcohol withdrawal syndrome, delirium tremens (also called the DTs or rum fits). After a longtime alcoholic stops drinking, the amount of REM sleep increases dramatically. On about the third day following alcohol withdrawal, the REM sleep rebound peaks and the proportion of REM sleep rises to nearly 100 percent. At that same time, hallucinations occur in waking. The major cause of death in acute DTs is instability of temperature control, and we know that such dyscontrol is seen in normal REM sleep and after sleep deprivation. It is as if when prolonged alcohol intake suppresses REM sleep the brain loses its thermoregulatory bearings.

SLEEP-ENHANCING MEDICATION

With the growth of sleep science, progress has been made toward finding a chemical that we can use to control sleep. But we still lack a naturally occurring sleep substance or a sedative that is completely free of side effects.

A new class of sedative is the commonly used benzodiazepines, which include some well-known tranquilizers. This class of drugs is effective but has several undesirable properties. First, long-term use may cause a subtle but pernicious addiction. As with other drug dependencies, the first sign of addiction may be a reoccurrence of the insomnia; this problem may be temporarily ameliorated by taking more drug only to have the insomnia worsen in the long run. The brain is becoming habituated—spoiled, as it were, by the drug. Moreover, because of the habituation, rebound insomnia often occurs when the person stops taking the benzodiazepine. If patient and doctor are unaware of the effect, they may want to go back to the drug or even to increase its dose. This simply postpones the day when the increasing debt will have to be paid. A final problem is that the drug is transformed by the body into other substances, whose cumulative build-up may interfere with alertness and attention during waking.

How do the benzodiazepines work to produce sedation? In the undrugged brain, inhibitory effects are often mediated by the neurotransmitter gamma amino butyric acid, or GABA. GABA is known to inhibit the activity of the dorsal raphe nucleus, whose output of serotonin is normally reduced during sleep. The benzodiazepines may thus work by increasing the GABAergic inhibition of the serotonergic system. In doing so, the sedative would help to suppress the neuronal activity that supports waking, arousal, and anxiety.

THE PARASOMNIAS

The parasomnias are sleep disorders characterized by the occurrence during sleep of phenomena normally seen only in waking. Parasomnias are of three main kinds: they involve either the respiratory system, the motor system, or the cognitive system. All are considered secondary sleep disorders, because the original dysfunction occurs in brainstem systems that control the three affected systems.

Sleep apnea

Men show markedly more respiratory variability in sleep than women and are more prone to respiratory problems. The tendency to snore, to stop breathing, and to experience collapsed and/or obstructed air passages in sleep is genetic and sex-linked. Here I discuss only one of the respiratory sleep disorders, sleep apnea.

Sleep apnea most often occurs in middle-aged obese males. In this disorder, the sleeper's fat-compromised airway collapses, interfering with breathing. Strained grunting may signal the sleeper's attempts to breathe, but his efforts will only worsen the attack. His skin may even turn blue as a result of declining oxygen levels. The individual begins breathing freely only when he awakens. Because over 300 attacks may occur within a few hours, the sleeper must awaken many times during the night.

Two sleep-related processes conspire to produce sleep apnea. First, the level of brain activation declines at sleep onset, particularly in the brainstem reticular formation, which contains the neural oscillator controlling respiration. The other process contributing to sleep apnea is the decline in muscle tone throughout the body during sleep, including the muscle tone of the tongue, the back of the mouth, the throat, and the airway itself (or pharynx). Even in normal sleepers, the airway collapses significantly at sleep onset, giving rise to the deep slow breathing pattern that says "he's asleep" on the sound track of a thriller when the victim's dark room appears. The two processes that conspire to impair respiration are diagrammed on the following page.

If and when the sleeper reaches REM sleep, two new factors may further aggravate the breathing problem. One is the now active suppression of muscle tone caused by motor inhibition, which further compromises the airway opening. The other is the respiratory irregularity that is normal in REM sleep— especially in association with the eye movements. Sudden increases in respiratory rate may actually decrease the efficiency of breathing because air is moved in and out of the upper airway but does not reach the depths of the lung where oxygen can enter the bloodstream. And breathing may also stop suddenly because breathing commands from respiratory neurons in the lower brainstem become disorganized by spread of the intermittent excitation from the pontine eye-movement centers.

Sleep apnea is often dramatically relieved by surgically by-passing the larynx or by weight loss. Sometimes the sleep apnea sufferer sleeps with a

This elephant seal dramatically illustrates the energy conservation function of sleep. Its metabolic rate is so low that it can afford to stop breathing in sleep. This animal's sleep apnea may be related to its adaptation to cold water, but we now know that some temporary respiratory arrests are quite normal during sleep in human males.

Top: *Because the respiratory center of the brainstem is a part of the reticular formation, commands to breathe are frequent and regular during waking when reticular neuronal activity is high (left). When reticular neuronal activity falls at sleep onset (right), the automatic brain commands to breathe become less frequent and, in many older men, stop altogether. This normal process, called central sleep apnea, may become pathological if it becomes so prolonged and frequent that it reduces the oxygen concentration in the blood, causing frequent awakenings.*

Bottom: *Reticular neurons control muscles that hold the airway open in waking (left). These neurons also diminish their output in sleep (right). As a consequence of this peripheral mechanism, the airway may completely collapse. The central and peripheral components may act together to prevent both sleep and breathing.*

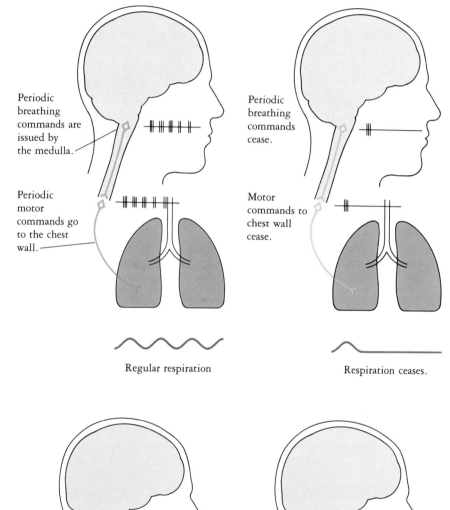

Periodic breathing commands are issued by the medulla.

Periodic motor commands go to the chest wall.

Regular respiration

Periodic breathing commands cease.

Motor commands to chest wall cease.

Respiration ceases.

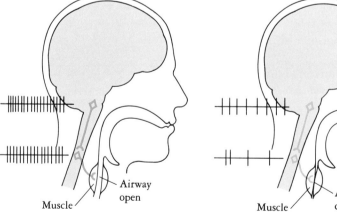

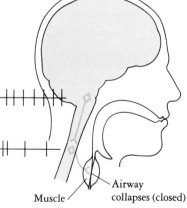

Muscle

Airway open

Ventilation continuous

Muscle

Airway collapses (closed)

Ventilation ceases

special mouthpiece that supports the tissues surrounding the airway in the back of the mouth. The airway pressure may be increased by connecting a pump to a mask; this procedure also helps maintain an open airway.

Motor disturbances of sleep

Sleep walking, sleep talking, and bruxism (or tooth grinding) are three of the most common motor problems in sleep. These automatic behaviors arise during Stages III and IV, and the EEG characteristics of these stages may persist throughout the entire somnambulistic episode. Because the motor activity may be complex and apparently purposeful, it is difficult to believe that the sleep walkers or sleep talkers are still asleep. These three sleep disorders are examples of the dissociation between cognition and behavior that occurs when the activation of the upper and lower parts of the brain is incomplete or uncoordinated in sleep. Because this dissociative process is also present in the enuresis (or bedwetting) and night terrors of children, the Canadian neurologist Roger Broughton has conceptualized all of these problems as "disorders of arousal" to signal the slowness and incompleteness of the upper brain's response to signals from below.

A similar dissociation occurs in nocturnal myoclonus, an affliction of older men. The automatic leg jerks of these men disturb their sleep enough to produce a feeling of fatigue the morning after but not enough to fully awaken them. The sleeper's wife, however, is usually fully aroused by her husband's unintended kicks.

Still another variation on the theme of dissociation is the sleep problem called the REM sleep behavior disorder. For reasons that are not yet clear, some sleepers, usually male, begin at about age fifty to lose the capacity to inhibit REM sleep motor commands. The result is the often comical but sometimes harmful acting out of dreamed movements. While dreaming of driving a car, the sleeper may move his arms as though turning the wheel. Or if he is playing football in a dream, he may run across the bedroom and tackle a chest of drawers. The dreamer may end up injuring himself—or his bed-partner should she happen to be in the way!

It is too soon to know either the causes or the course of this newly described sleep disorder, but some afflicted individuals have been found to have a neurological disease of the brainstem. Such findings suggest a relationship to the similar condition, REM sleep without atonia, that can be experimentally produced in cats. The one-to-one relationship between the observed motor behavior and the dreamed experience of it gives strong support to the activation-synthesis theory of dreaming discussed in Chapter 7.

The terrifying visions that may persist even after awakening from a frightening dream are represented in this detail from the Swiss artist Ferdinand Hodler's painting called The Night *(c. 1890). Given the visual and emotional intensity of REM sleep dreams, it is not surprising that these processes sometimes spill over into subsequent waking.*

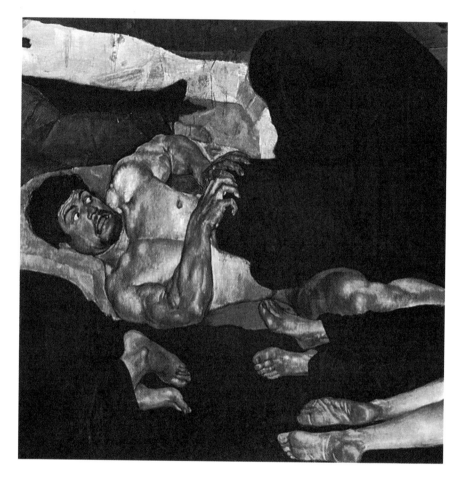

Nightmares and night terrors

In and around sleep some people may experience thoughts and feelings that are so similar to those of mental illness that they and their families become concerned. Because the brain is rapidly changing its electrical and biochemical state throughout the night—and because so many of those changes require the precise timing of sometimes opposed brain forces—it is not surprising that emotion and altered perception and thinking are sometimes so strongly triggered that the sleeper awakens. These phenomena are so common at certain ages that they are practically normal.

A child experiencing a night terror will be aroused, only partly awake and in panic, from Stage III or Stage IV of non-REM sleep. His heart will beat quickly, and respiration will be rapid. Although able to move normally and even to listen and converse, the child may terrify his parents by experiencing hallucinations. Night terrors do not have serious consequences, however, and the child outgrows them.

Nightmares occur at all ages. Some are precipitated from non-REM sleep, in which case they are characterized by pure fear without visual imagery, and others from REM sleep, when they take the form of vivid and frightening dreams. In both types, dramatic heart palpitations, increases in blood pressure, and drenching sweats may occur. The non-REM nightmares can sometimes be stopped by means of benzodiazepines which suppress Stages III and IV of non-REM sleep. The REM nightmares are not pharmocologically treatable, and the interruption of sleep may give rise to insomnia.

Whereas night terrors and nightmares are sleep phenomena that are not related to previous trauma, in posttraumatic stress disorder sleep is troubled specifically by the recurrent experience of a previous traumatic episode. This disorder is found among war veterans, concentration camp survivors, or individuals otherwise exposed to extreme physical or psychological violence. By day, these individuals are fearful or guilt-ridden. They may experience flashbacks even when awake, and especially under hypnosis.

Typically, the nocturnal reliving of such experiences produces anxiety of monumental proportions. This anxiety resembles and often surpasses that of night terrors. The affected person may partly awaken and act out episodes of the traumatic experience, sometimes unwittingly attacking a family member. In a sense, the posttraumatic stress disorder represents the high end of the spectrum of breakthrough of motor and mental activity in sleep. It combines in its nocturnal manifestations sleep walking, sleep talking, and night terrors. The experiences are often so vivid that the affected individual identifies them with dreaming. However, these phenomena, like most of the other motor and mental parasomnias I have discussed, normally arise from non-REM sleep and are thus an important exception to the rule that hallucinoid dreaming is associated with REM sleep. Like other non-REM phenomena, they may be considered to be manifestations of physiological instability during phase transitions or exaggerations of the normal capacity to have motor activity during this phase of sleep. Similarly, the impaired cognitive functions resemble those that occur on arousal from deep non-REM sleep.

The treatment of posttraumatic stress disorder usually involves a combination of cathartic psychotherapy, hypnotic suggestion, and attempts to suppress non-REM sleep through benzodiazepines.

By focusing on what causes sleep disorders and how they might be alleviated, this chapter has attempted to integrate sleep disorders into the unified conceptual framework provided by the reciprocal interaction hypothesis. I will now turn my attention to the functional gains that the brain may derive from indulging in this complex and risky behavior called sleep.

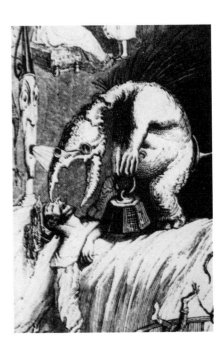

The pure terror and strong sense of oppression that may accompany arousal from non-REM sleep are symbolically represented as a giant crustacean lowering a heavy weight onto the sleeper's chest in this print by the French graphic artist Jean Ignace Isidore Gérard, called "Grandville." Such classic "nightmare" experiences are thus psychologically and physiologically distinct from the bad dreams represented by Hodler.

C H A P T E R

9

The Functions of Sleep

Sleep is risky. By abandoning temperature control, we make ourselves vulnerable to being frozen or cooked; by abandoning vigilance, we expose ourselves to surprise attack; by abandoning controlled consciousness, we risk committing errors of perception, logic, and judgment. These are the dangers of entirely normal sleep. In addition, the system that organizes sleep may make errors that emerge as sometimes fatal disorders.

Why does nature take such risks? Or, to put it more scientifically, how do the benefits of sleep outweigh the risks? As they must have, given the evolutionary success of animals that sleep. Two complementary answers have already emerged in the course of this book. First, by lowering our metabolic rate in sleep, we conserve energy, at the same time reducing the risk of thermal disequilibrium during the coldest part of the day. Second, because new learning processes are inactivated, during sleep we can reorganize and more efficiently store the information already in the brain.

After a night's deep sleep, we awaken. How has our sleep affected us? Most scientists are convinced that both energy conservation and information processing are served by sleep, but we do not yet have an objective physiological measure of our subjective sense of restoration and improved mental alertness after a night of sound sleep. The mystery of sleep function is still impenetrable.

Energy conservation theories of sleep are called homeostatic because they concern the maintenance of thermal equilibrium. The low metabolic demands and decreased body temperature of sleep are ways for this arctic fox to save internal energy. Insulated by its fine fur coat, the fox further reduces heat loss by assuming a curled posture (which diminishes radiant surface area) on a bed of snow (which insulates the fox from the ground).

After these two functions have been served by sleep, we are better prepared to handle the demands of our waking hours. By day our bodies are able to meet a widely fluctuating set of energy demands, as they generate more or less energy depending on outside temperature, the exertion required to obtain food and shelter, and other factors. At the same time, our brains are better able to both accrue data and guide our behavior.

More generally, we can say that sleep serves the functions of both homeostasis (or constancy) and heteroplasticity (or change), which are required for successful adaptation. Homeostasis refers to the tendency of physiological systems to resist change and maintain a constant set of internal conditions. Heteroplasticity refers to the capacity to change in response to new circumstances. While homeostatic systems are mainly concerned with energy regulation and heteroplastic systems are mainly concerned with information processing, the two functions can overlap. In this chapter, I consider a few of the possible homeostatic and heteroplastic functions of sleep.

One function that we would like to hold constant is our level of alertness, but it fluctuates despite our best intentions. And sometimes it fails dramatically. When this happens, we know we need a good night of sleep, and sleep usually helps to restore alertness. How could sleep serve this particular homeostatic purpose?

REPLENISHMENT OF NEUROTRANSMITTERS

Although most neurons decrease activity slightly in sleep, a small minority actually cease firing altogether. And as we saw in Chapter 6, this cessation begins in non-REM sleep and is complete in REM. Because these REM-off cells are norepinephrine- and serotonin-releasing neurons of the locus coeruleus and the raphe nuclei, we wonder what is the significance of the fact that the population of neurons with this REM-off pattern is aminergic. Might this arrest of firing be in the service of conserving aminergic neurotransmitter?

Let us assume that, over the course of the day, the average number of molecules of neurotransmitter released per discharge declines as the result of neurotransmitter depletion. The proportion of signals that are successfully ferried across the synapse would decline correspondingly. Now, by *not* firing in REM sleep, the cells of the locus coeruleus and raphe nuclei, which have

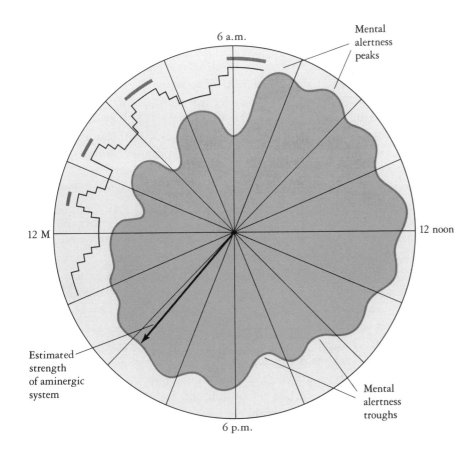

The relative efficacy of the aminergic neuronal system over a 24-hour period is schematically represented by the distance of the orange-red curve from the center of the circle. The sixteen subdivisions of the circle represent the 90-minute basic rest-activity cycle postulated by Kleitman; they are seen most dramatically during sleep as the periodic alteration of REM and non-REM phases. Because the output of the aminergic system is at a low ebb during sleep, neurotransmitter may be conserved. As a result, we experience maximal mental efficiency in the morning. Thereafter alertness declines steadily (though with superimposed 90-minute ripples) over the course of the day until sleep again ensues.

been monotonously and relentlessly active all day long, are spared from expending norepinephrine and serotonin. If these aminergic neurons continue to synthesize new transmitter during sleep, and the neurotransmitter, in turn, continues to be sent from the cells' nuclei to the axonal endings throughout the cortex, then a night's sleep that is rich in REM will find the neurons with more neurotransmitter available for release when morning arrives.

What a wonderful way to explain those striking cognitive lapses that occur more and more frequently as the day wears on and that can become quite alarming after nightfall! And having slept, don't we really feel mentally sharper—not just more alert, but *better* at cognitive tasks like writing and problem solving? Of course, we know there are evening people (owls) as well as morning people (larks), and this may mean the theory is somehow wrong or incomplete—or it may mean that circadian factors play on the system in ways that overcome or displace the effect of neurotransmitter depletion and replenishment.

The fact is that the aminergic neurons, thought to be crucial to attentive learning and memory, *do* rest in sleep and do so most decidedly in the REM phase. And that has got to be of some functional interest.

Further evidence for the homeostatic function of sleep comes from the startling findings of Rechtschaffen on the ultimately fatal effect of sleep deprivation upon energy and temperature regulation in rats. As detailed in Chapter 5 and shown in the graph on the facing page, the dying rats show a paradoxical weight loss despite increased food intake, both occurring at the same time that temperature control is lost. It is as if the animals were literally burning up. To make the analogy more graphic, imagine two wood stoves that have different draft controls: the opening in one draft control can be adjusted to any width, but the other control has only two settings: closed and wide open. The same amount of fuel is placed in each stove and the fire started; both draft controls are wide open. Off goes the fire. But while the one stove can now be adjusted to produce a long steady heat supply, the all-or-none stove must either shut off (and die) or burn rapidly to completion (and die).

Sleep-deprived animals are like the wood stove without draft control; their heat goes whooshing up the chimney, and they soon need more fuel to stay alive. The sleep-sated animal has smoothly modulated control; energy flow is regulated efficiently during the day, and the damper is shut almost tight at night. Thus, if we do not sleep, we lose the capacity to modulate energy flow with maximal efficiency. Our feeling of fatigue is our subjective awareness of this shift toward energy consumption. We even say, "I need some sleep," as if sleep were somehow a foodstuff or a fuel. But what we actually need is not more fuel, but to rebuild the brain system that controls the consumption of fuel.

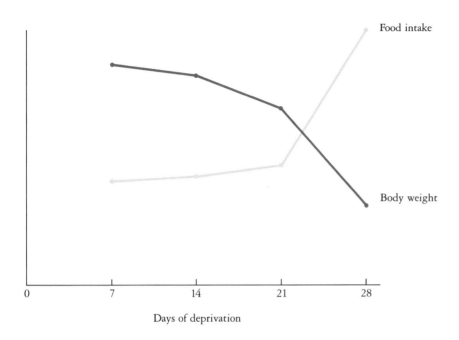

Food intake

Body weight

0   7   14   21   28

Days of deprivation

After three weeks in the apparatus shown on page 114, sleep-deprived rats begin to show contrary changes in metabolism (red line) and food intake (beige line). The failure to maintain body weight (or body temperature) despite massive increases in food intake suggests that sleep may serve to restore the efficiency of brain mechanisms that regulate energy.

HETEROPLASTIC THEORIES

Heteroplastic theories of sleep function fall into two main categories. Some emphasize the role of sleep in the unfolding of genetically determined programs; these are the developmental theories. Other hypotheses, called learning theories, emphasize sleep's role in bringing about environmentally engendered modifications in those programs and its role in subsequent maintenance of these programs in the animal's repertoire of behavior.

In this book, I have already touched on many general considerations that give heteroplasticity theories their appeal. The obvious benefits of sleep to sensorimotor, emotional, and intellectual functioning go beyond the level of mere recovery. We often do *better* after sleep, suggesting that we have not just lost our tiredness but have actually elaborated new ability to be attentive and to remember. I have to say at the outset, however, that none of the hypothesized heteroplastic functions is as yet well established. Although some of the theories may today seem mutually incompatible, they could turn out to be mutually supporting in the light of tomorrow's science. Scientists can now only outline an agenda, a program for investigating these functions, and in recounting some theories of possible sleep function, I will be understandably opportunistic, looking mainly where the light is.

DEVELOPMENTAL THEORIES

All current developmental theories of sleep function focus on REM sleep, since it is such a prominent behavior in utero and in early infancy. In these theories, the activation that occurs in REM sleep provides an opportunity for the brain to practice future behavior. The brain thereby increases the ability of neurons and circuits to function competently before the organism is called upon to use them.

REM sleep in utero has several features that are in keeping with this hypothesis. First, REM sleep is reliable. The system must activate in the same way on each occasion and in each individual if it is to functionally extend a genetic blueprint. It thus makes sense to have the activation emanate from the brainstem, as REM sleep does, since the brainstem is the oldest part of the brain and millions of years of evolution have perfected its workings. Second, the developing organism is guaranteed plenty of time in the activation mode,

REM sleep may play a role in developing the fetal brain. During REM sleep, the brainstem would signal the arm to move and at the same time activate sensory and motor systems in the cortex. The brain would then be able to build and refine the activated circuits.

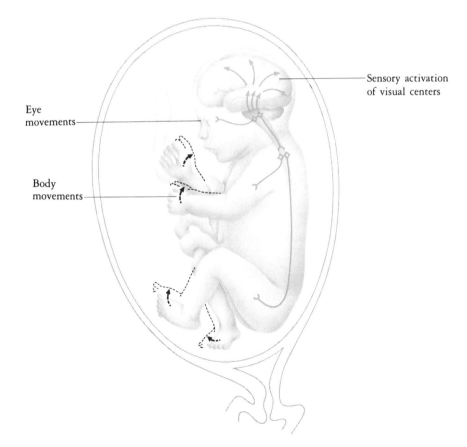

Sensory activation
of visual centers

Eye
movements

Body
movements

Marsupials are born immature and develop in pouches or nests where they spend most of their time asleep. Huddled together for warmth, these eight-week-old Australian brown antechinus could help answer questions about how early in mammalian development REM sleep actually occurs and what role it plays in establishing the functional architecture of the brain.

since the fetus spends as much as 80 percent of its time in REM sleep. Third, REM sleep activates the PGO circuitry, which provides repetitive, stereotyped internal stimulations that could be used to develop motor and sensory systems.

I have already discussed in Chapter 4 one theory that gives REM sleep a role in the development of brain function. According to Howard Roffwarg, during REM sleep the brainstem provides stimulation in the form of PGO waves to the visual system in the cortex. In responding to the signals from the brainstem, the cortex builds up and refines its internal circuits. The Roffwarg theory emphasizes activation of sensory systems, because the visual system is the target of so much rhythmic stimulation during REM sleep. Since the congenitally blind do not "see" in their dreams, it is obviously not enough to have internal stimulation; the external world must provide formed visual inputs, and visual memory must then be used to construct dream imagery.

We know, from research on the development of binocular vision, that the visual system of the cortex is at least in part shaped by the images an infant views in his first year of life. Prenatal REM sleep could prepare for this critical early experience by making visual system neurons ready to respond to formed images. It is as if the newborn's nervous system has two similar activation modes—waking and REM sleep—and the two modes reinforce each other's efforts to develop the visual system's function.

Fetal motor systems may also be developed during REM sleep. All that an animal does, including the breathing and feeding necessary for survival, depends upon organized movement. REM sleep provides an internal activation of the brainstem motor systems controlling these vital rhythmic movements. We "know" how to breath and to swallow when we are born. How can this be? Have we already "learned" these acts during REM sleep in utero?

Direct observation of fetal lambs through Plexiglas windows implanted in the uterine wall suggests that the answer is yes: despite the absence of air to breathe, breathing movements of the chest wall are seen in REM sleep. When we recall that breast-feeding babies have sucking movements that are rhythmically linked to the REM sleep eye movements, we see another way in which the brainstem may organize significant behaviors by means of its own activation during sleep.

MAINTENANCE THEORIES

After birth, brain development begins to be influenced by our experiences as well as by genetic programming. These experiences make their impact on the brain through the process of learning. But another process that becomes important at this point is that of *not* forgetting. How can a brain system retain its acquired competence without practice? Whether the brain is carrying out a reflex, a set of reflexes, or a chain of acts, it has to retain its capacity to direct performance for years, even if occasions for performance do not arise.

In the case of eye movements, not a minute goes by in waking without our making hundreds of them. So it is hard to see why we would need to practice them in our sleep. But the fact that eye movements do occur in sleep led Ralph Berger of the University of California at Santa Cruz to suggest that the visual perception of depth might be facilitated by REM sleep. Attempts to test this idea using sleep deprivation have thus far not been successful, but sleep deprivation experiments are difficult to perform effectively without severely disrupting the animal's metabolic control systems.

Other behaviors, like those expressing sexual desire, fright, and aggression, are not so constantly elicited in waking. Thus, Michel Jouvet has proposed that REM sleep provides an opportunity for rehearsing less common behaviors. This idea is supported by observations of cats whose brainstems have been damaged so as to prevent motor inhibition. We can actually see these cats act out instinctual behavior in REM sleep. And our dreams are full of instinctual acts.

I like Jouvet's idea because it fits with the proposition that in dreaming the brain is reinforcing critical information stores. I wonder whether even cats who had never fought (or been attacked) or had never been exposed to sexual activity (or a partner) would evince REM sleep aggression and sexual behavior. This is an easily testable hypothesis: one need only raise cats in isolation from other cats to find the answer. Another "maintenance" theory of sleep will be discussed in the next section because its emphasis is on the reinforcement of memory.

LEARNING THEORIES

We do not learn anything new in sleep; that is, we cannot acquire new information in that state. Playing a language learning tape under your pillow is *not* a good way to become fluent in French. The reason is that even though some stimuli may get into the sleeping brain, they are not retained unless arousal occurs. But we have all had the experience of waking up with a problem solved. Furthermore, although the changes are quite small, animals do show increases in REM sleep while they are being trained. And REM sleep deprivation impedes new learning.

One way that REM sleep could aid the learning process is by reinforcing memory. While scientists have long suspected that memories are stored in the brain, they have only recently shown that nerve cells actually change their responses in the light of prior experience. In the simplest form of learning, called sensitization, cells respond briskly to stimuli if they have recently been exposed to those stimuli. Conversely, too frequent exposure leads the response to become attenuated.

Most of these neuronal forms of learning have been studied in simple animals (like snails) or in brain slices (like the hippocampus), and they have all been shown to require such neurotransmitters as norepinephrine, serotonin, and acetylcholine. How the more complex human brain—with its myriad neurons and multiple circuits—organizes memory is still unknown. Most neuroscientists now believe that although some parts of the brain are critical for some kinds of learning, memory is ultimately distributed throughout the brain. Experience is encoded as an increase in the synaptic strength of whatever circuits are engaged during exposure to new information. (When synaptic strength increases, fewer impulses at the presynaptic ending are necessary for the postsynaptic neuron to fire.) Many scientists suspect that for any information to be permanently stored in the brain, it has to be converted from a constellation of changes in electrical activity into the protein structure within neurons.

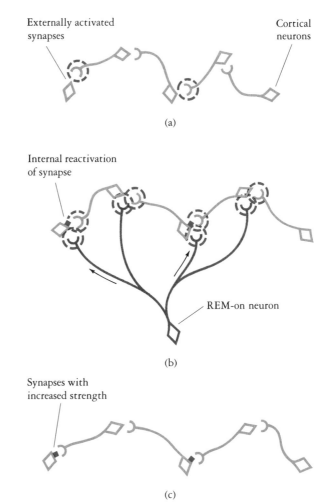

According to one theory, sleep may encourage memory by enhancing changes in the synaptic strength of those specific neural networks that are activated by the experience to be remembered. (a) New information acquired during waking experience temporarily activates different sets of synapses, now called "hot spots" (dashed circles). (b) During REM sleep, all synapses in the brain are automatically activated. As the cholinergic system (purple) interacts with the hot spots, those synapses change their character, and (c) the information stored in the hot spots is converted into a more permanent form (red boxes).

This new concept of memory implies that a specific memory could be enhanced simply by reactivating the circuits whose synaptic strengths had been altered by previous experience. Since such activation is present in REM sleep, we have reason to suspect that sleep—outward appearances to the contrary—may facilitate learning and memory. The diagram on the facing page illustrates this process.

REM sleep activation could help us to remember not only recently learned information but also information acquired long ago. The microbiologist Bernard Davis suggested that our memory must be constantly renewed to avoid decay. One way to accomplish this renewal is to relearn the material in waking, and we do this with many of our skills—for example, our native language. But how do we manage to retain a foreign language that we do not have occasion to practice?

Davis reasoned that acquired data is stored in unstable proteins, which would run down if not refreshed. According to Davis, the data stored in our brains is refreshed by the automatic pattern of activation seen in REM sleep. As signals from the brainstem activate higher areas of the cortex, they would reinforce all the data in the brain according to its current synaptic strength. Thus, highest priority would be given to new data and a lower priority to old data that had not been used.

Drawing an analogy with the immune system (where brief reexposure to an antigen is enough to provoke a powerful antibody response), Davis thinks that maintaining even a limited ability to recognize objects after many years would be helped by automatic brain activation. He proposes that the firing of neurons in patterns that reproduce the original input could reasonably achieve this goal. To make this process work properly, the system might block both input and output, and then activate groups of neurons in such a way as to assure many different patterns of neurons are stimulated. According to Davis, the stimulation would have to scan the entire brain to be effective. This is exactly what appears to happen in REM sleep, and our dreams are the subjective experience of this mnemonic reiteration. We seem to be rememorizing our memory every night of our lives.

SLEEPING TO FORGET

Francis Crick and Graeme Mitchison have proposed that REM sleep serves to remove undesirable data from memory. They suggest that just as it is important to reinforce certain associations—and the networks of neurons encoding them—it is equally crucial to weaken others. Suppose, for example, I had to remember every detail of my Friday experience and of my Saturday experience and of every day thereafter. My brain-mind would soon be stuck in an endless loop of trivial memory. I would resemble the idiot savant whose head is so full of tabular data—like the Toledo street directory—that more relevant information simply cannot be processed.

The psychological fact that dreams are so difficult to remember suggests to Crick and Mitchison that the process might have been designed to erase, rather than strengthen, certain memories. "We dream in order to forget," they say. For an explanation of the mechanism causing memory erasure, they look to the interaction between the PGO waves of REM sleep and oscillations between cortical neurons. Any highly interconnected system like the cerebral cortex is liable to become self-exciting in an uncontrolled way and to develop oscillations. These oscillations might result in epileptic seizures or in persistent but unproductive ideas, like obsessions. By innocuously triggering the potentially pathogenic oscillations, the PGO system actually *reduces* the synaptic strength between some neurons, causing information to be lost.

Although the Davis theory emphasizes how sleep helps us to remember and the Crick–Mitchison theory how sleep helps us to forget, the two theories are not incompatible, either theoretically or empirically. The brain is obviously always doing two things at once. Memory is a two-way street—some items are going in, others are going out, and a double mechanism must always be at work.

BRAIN CHEMISTRY AND THE FUNCTION OF SLEEP

In this chapter, I have suggested that sleep affords ideal conditions for our brains to restore two important functions: energy management and information management. The conditions are ideal because no new information is being taken in (since input is blocked), because the process itself is not recorded (since dreams are forgotten), and because the process has no behavioral effects (since there is no motor output). In both energy and information management, there is a loss during sleep that is compensated by a gain during waking:

Energy: We don't control temperature now in order to control it better later.

Information: We don't take in any new data now in order to conserve what has already been acquired and to do better acquiring later.

Both energy and information management are made possible by the arrest of aminergic neuronal firing in REM sleep.

1. By mediating changes in the generation and flow of body heat, aminergic neurons participate directly in temperature control. We may thus be confident that there is at least a significant correlation, if not a causal connection, between sleep, aminergic synaptic efficacy, and thermoregulatory control.

2. Aminergic neurons participate directly in information acquisition, and without them memory is impossible. (It is known that cells do not keep a record of a cholinergic stimulus unless aminergic interneurons are coactivated.) Thus, there is a strong correlation and, I contend, a causal connection between sleep, aminergic synaptic efficacy, and information control.

Because the aminergic neuronal system stops firing, the disinhibited cholinergic system acts unopposed in REM sleep. I would like to propose that in REM sleep the brain reinforces its genetically determined programs and integrates new experiential data by means of intense cholinergic stimulation. This proposal is inspired by the work of Swiss physiologist Walter R. Hess. Long before the cellular and molecular basis of this kind of shift in central brain neurotransmitter ratio had been proposed, he had suggested a contrast between the sympathetic (aminergic) and parasympathetic (cholinergic) divisions of the autonomic nervous system, whose physiology was then well-established in the periphery of the body. The former was concerned with action, and the latter was concerned with rest and recovery. These divisions have an obvious parallel in my theory, which includes information acquisition (on the aminergic side) and information conservation (on the cholinergic side).

Speculation regarding the functional significance of changes in neurotransmitter ratio during sleep is supported by neurobiological data from several independent sources. First, the PGO burst cells of the peribrachial pons—which fire clusters of spikes in REM sleep—have tentatively been identified as cholinergic because they coincide with one of the two known groups of such cells in the brainstem. The implication that the thalamus and cortex are receiving pulses of acetylcholine during REM sleep has since been confirmed by Steriade and his team in Quebec. This fact possibly explains the previously reported observation of the Montreal physiologist Herbert Jasper that acetylcholine release from the cortex is as high in REM sleep as during waking.

The rise in acetylcholine release combined with the decrease in aminergic transmitter release means that the aminergic/cholinergic ratio of the cortex must change by a hundred to a thousandfold (from about 1 to at least .01)! What metabolic operations inside a neuron could cause such chemical shifts?

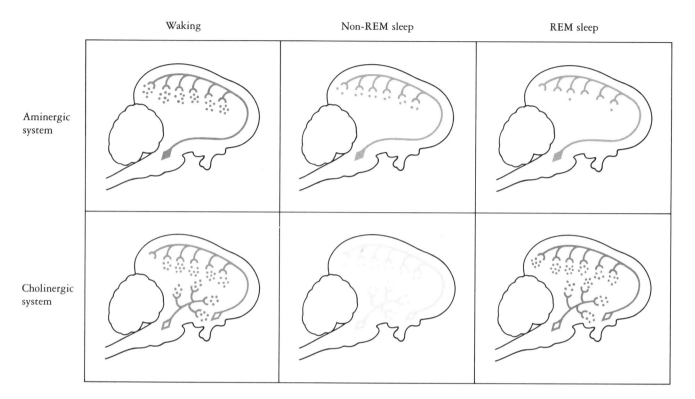

| | Waking | Non-REM sleep | REM sleep |

Aminergic system

Cholinergic system

During waking both the aminergic (red) and cholinergic (purple) neurons are active, and the brain is bathed in high and balanced levels of both kinds of neurotransmitter (dots). In non-REM sleep, the balance is retained, but the levels of both decline (fewer dots). Alertness also declines, but the logical and thoughtlike nature of mental activity does not change. In REM sleep, the chemical balance of the brain shifts radically as aminergic neurotransmitter levels plummet to their nadir and cholinergic neurotransmitter levels rise again to waking levels. The effects of this shift are seen in the unique formal features of dreaming, including bizarre events, illogical explanations, and poor recall.

The work of Ira Black, a neurobiologist at Cornell University, provides some hints. Black's experiments focused on the superior cervical ganglion, a nubbin of nerve tissue in the neck that receives presynaptic fibers from the brain and projects postsynaptic fibers to the pupil of the eye. When stimulated, this ganglion causes the pupils to dilate. Black found that when he stimulated the ganglion, the number of neurotransmitter-synthesizing enzymes inside the cells rose fiftyfold. Hence, aminergic neurons change their own rate of neurotransmitter synthesis according to their level of use. Since the enzymes are synthesized by the cell itself, Black's work demonstrated a direct and powerful link between the electrical activity level of nerve cells and their biochemistry. Since brain cell activity rises and falls dramatically during sleep, we may safely infer proportional changes in brain biochemistry.

The most critical demand of a new speculative theory is that it can be tested. Our ability to manipulate REM sleep in controlled experiments may now be used to find out precisely how brain chemistry is changed. For example, REM sleep could be prevented (by deprivation) or increased (by cholinergic stimulation) and the capacity of the brain to synthesize norepinephrine under these conditions measured. We would predict a progressive fall in biosynthesis as REM deprivation becomes extreme and a rise to a peak during prolonged REM sleep.

Such experiments could advance our knowledge of sleep mechanism and function to the level of molecular biology. We might achieve more realistic molecular models of both the circadian and ultradian rhythm generators, whose genetic basis is strong but whose mechanism is still obscure. And we might understand how sleep engineers its energy and information conservation functions in specific molecular terms. We should not be surprised to learn that both our homeostatic and heteroplastic goals were served by common molecular processes in sleep.

Beneath its cloak of outward calm, sleep conceals a rich array of dynamic processes. Despite our apparent inactivity, physiological recordings reveal continuous brain rhythms of remarkable complexity and reassuring regularity. And although our own consciousness and memory seem to be obliterated, psychological studies disclose mental processes whose changes throughout the night reflect, in surprisingly precise ways, the constantly shifting balance between groups of neurons in the brain. The variations in sleep and its underlying brain mechanisms within and across species suggest a twofold purpose: to conserve energy and to organize information. Learning exactly how these functions are served is now the major agenda of sleep science.

Bibliography

J. Allan Hobson. *The Dreaming Brain*. New York: Basic Books, 1988. 319 pp.

An account of the development of sleep and dream research from the mid-nineteenth century to the present, with special emphasis on the activation-synthesis hypothesis of dreaming.

———. *The Chemistry of Conscious States*. New York: Little, Brown & Co., 1994.

A physiological model of conscious experience is sketched and illustrated by examples taken from everyday experience and from clinical data.

J. Allan Hobson et al. Dream Consciousness: A Neurocognitive Approach. *Consciousness and Cognition*, 1995. 3:1–128.

Nathaniel Kleitman. *Sleep and Wakefulness.* Revised and enlarged edition. University of Chicago Press, 1963.

An encyclopedic reference text describing Kleitman's own seminal views and citing the data of thousands of studies from the premodern period.

M. H. Kryger, T. Roth, and W. C. Dement. *The Principles and Practice of Sleep Medicine.* Philadelpia: W. B. Saunders, 1994.

A concise and well-illustrated overview of sleep disorders written primarily for physicians but accessible to the motivated lay reader.

Mircea Steriade and R. W. McCarley. *Brainstem Control of Wake-Sleep States.* New York: Plenum, 1990.

A detailed and extensively documented technical monograph describing the modern period of sleep neurobiology, with a focus on cellular and molecular mechanisms.

Sources of the Illustrations

intensified Video Microscopy and Flow Cytometry," *The Journal of Cell Biology,* vol. 100, May 1985, fig. 3.

page 33
Adapted from Jurgen Aschoff, *Science,* vol. 148, 1965, p. 1427.

page 36
left, Robert A. Tyrrell
right, Gregory Florant

page 42
Nathaniel Kleitman

page 46
Jonathan Scott/Planet Earth Pictures

page 49
C. A. Henley/Larus Natural History Photographs

page 51
Truett Allison

page 54
John Shaw

page 55
bottom, Frans Lanting

page 56
top, Edward Tauber

page 57
bottom, Frans Lanting

page 59
C. A. Henley/Larus Natural History Photographs

page 61
Ernest Hartmann, *The Biology of*

Dreaming. Springfield, Ill., Charles C Thomas, 1967.

page 62
Frans Lanting

page 63
top, William Dement
bottom, Jonathan Scott/Planet Earth Pictures

page 65
Dwight Kuhn

page 66
Richard Wrangham/Anthro-Photo File

page 70
Ted Spagna

page 72
Adapted from Howard Roffwarg, *Science,* vol. 152, 1966.

page 74
Jason Birnholz

page 79
Anthony Joyce/Planet Earth Pictures

pages 85, 87, 89, 91
right, Ted Spagna

page 92
Roger La Fosse

pages 96, 99, 102, 103
Ted Spagna

page 104
Frans Lanting

page 107
John Shaw

page 110
Stephen Dalton/NHPA

page 116
J. Allan Hobson

page 130
J. Allan Hobson

page 131
top, J. Allan Hobson
bottom left, Ragnhild Karlstrom
bottom right, J. Allan Hobson

page 133
top and bottom, Reinhard Grzanna

page 136
Marcel Mesulam

page 138
top and bottom, James Quattrochi

page 170
Ted Spagna

page 183
Frans Lanting

page 188
Al Rubin

page 190
Fred Bruemmer

page 195
C. A. Henley/Larus Natural History Photographs

Index

Selected hardcover books in the
Scientific American Library series:

A JOURNEY INTO GRAVITY AND SPACETIME
by John Archibald Wheeler

ATOMS, ELECTRONS, AND CHANGE
by P.W. Atkins

DIVERSITY AND THE TROPICAL RAINFOREST
by John Terborgh

STARS
by James B. Kaler

GENES AND THE BIOLOGY OF CANCER
by Harold Varmus and Robert A. Weinberg

MOLECULES AND MENTAL ILLNESS
by Samuel H. Barondes

EXPLORING PLANETARY WORLDS
by David Morrison

EARTHQUAKES AND GEOLOGICAL
DISCOVERY
by Bruce A. Bolt

THE ORIGIN OF MODERN HUMANS
by Roger Lewin

THE EVOLVING COAST
by Richard A. Davis, Jr.

THE LIFE PROCESSES OF PLANTS
by Arthur W. Galston

IMAGES OF MIND
by Michael I. Posner and Marcus E. Raichle

THE ANIMAL MIND
by James L. Gould and Carol Grant Gould

MATHEMATICS: THE SCIENCE OF PATTERNS
by Keith Devlin

A SHORT HISTORY OF THE UNIVERSE
by Joseph Silk

THE EMERGENCE OF AGRICULTURE
by Bruce D. Smith

ATMOSPHERE, CLIMATE, AND CHANGE
by Thomas E. Graedel and Paul J. Crutzen

AGING: A NATURAL HISTORY
by Robert E. Ricklefs and Caleb E. Finch

INVESTIGATING DISEASE PATTERNS:
THE SCIENCE OF EPIDEMIOLOGY
by Paul D. Stolley and Tamar Lasky

Other Scientific American Library books
now available in paperback:

POWERS OF TEN
by Philip and Phylis Morrison and the Office of
Charles and Ray Eames

THE DISCOVERY OF SUBATOMIC PARTICLES
by Steven Weinberg

THE SCIENCE OF MUSICAL SOUND
by John R. Pierce

THE SECOND LAW
by P.W. Atkins

MOLECULES
by P.W. Atkins

THE NEW ARCHAEOLOGY AND THE ANCIENT
MAYA
by Jeremy A. Sabloff

THE HONEY BEE
by James L. Gould and Carol Grant Gould

EYE, BRAIN, AND VISION
by David H. Hubel

PERCEPTION
by Irvin Rock

FROM QUARKS TO THE COSMOS
by Leon M. Lederman and David N. Schramm

HUMAN DIVERSITY
by Richard Lewontin

If you would like to purchase additional volumes
in the Scientific American Library, please send your order to:

Scientific American Library
P.O. Box 646
Holmes, PA 19043-9946